Prehension

Prehension

The Hand and the Emergence of Humanity

Colin McGinn

The MIT Press
Cambridge, Massachusetts
London, England

MIT Press books may be purchased at special quantity discounts for business or sales promotional use. For information, please email special_sales@mitpress.mit.edu.

This book was set in Stone by the MIT Press. Printed and bound in the United States of America.

Library of Congress Cataloging-in-Publication Data

McGinn, Colin, 1950–
Prehension : the hand and the emergence of humanity / Colin McGinn.
     pages   cm
Includes bibliographical references and index.
ISBN 978-0-262-02932-2 (hardcover : alk. paper)
1. Human evolution. 2. Hand—Evolution. I. Title.
GN281.M395 2015
599.93′8—dc23

                                                                        2014049819

10   9   8   7   6   5   4   3   2   1

Contents

Preface

I expect this book to be controversial. Of the six readers who reviewed it for the MIT Press, half hated it and half were enthusiastic. The former group consisted of people with a scientific background, the latter group of philosophers. It seems reasonable to predict a similar division (not to say divisiveness) among readers going forward. I don't think this in itself is a bad thing. The book is certainly quirky, speculative, and unorthodox. As they say, it pushes the envelope. I give my imagination a freer rein than usual—too much for some people, perhaps.

I am not an expert in human evolution or anatomy or archaeology. I have never even touched a human fossil. My knowledge of these fields is strictly that of a layman. What I have done is combine an amateur knowledge of the science with a philosophical slant (in which I do claim some expertise). I keep the science basic and general, though I hope not naïve. I use the science as a springboard for speculation. This is certainly not a textbook on human evolution or hand physiology. It is an essay in philosophical anthropology—general reflections on the nature of humankind from an evolutionary perspective.

I have pitched it at roughly the level of Richard Dawkins's *The Selfish Gene*. That book seeks to conceptualize, organize, and dramatize the scientific facts, and is philosophical to that degree. It is also written for the general reader, not the specialist (though it is certainly of interest to a specialist). I have attempted something similar here, though my efforts are a pale shadow of that brilliant book. This book is directed to general readers and philosophers (though scientists more given to speculation might get something out of it). My procedure tends to be "rational reconstruction": seeing how things might have been, seeing what is intelligible. Empirical verification is another matter, a very difficult one given the distances of time and lack of concrete evidence. Yes, I am telling stories, though stories rooted in science, and intended seriously.

The human hand plays a central role in the story I tell. One of my aims here is to make us think about something we tend to take for granted: the importance of the hand in nearly all of our endeavors. I even wax a bit poetic about it in places. I approach the hand from multiple directions—anatomically, functionally, emotionally, cognitively, artistically, and philosophically. This is, in part, a hymn to the hand. Touching, not seeing, gripping, not witnessing: hence my title. This perspective is by no means original to me: many thinkers have recognized the importance of the hand and have written eloquently about it. I am simply continuing that tradition, adding a point here and there.

Human evolution is still shrouded in mystery in many ways, and there may even be limits to what we can expect to explain. Still, a picture emerges that is both chastening and uplifting— a possible way things could have been, and may actually have been. I certainly don't claim to have the full picture. But even the sketch is worth making, with all due modesty. What I seek to provide here is a philosopher's take on the sketch that has emerged since Darwin's *The Descent of Man*. It is a mixture of biology, anthropology, psychology, analytical philosophy, existential philosophy, sheer speculation, and utter amazement. I hope the reader will forgive my sometimes exuberant style and neglect of some of the scientific details (when I think the details don't affect the big picture). This is less a scholarly tome than a dithyramb to the powers of evolution in general and the human hand in particular. And it is not afraid to go out on a limb (an apt metaphor, given the emphasis on trees).

Colin McGinn
September 2014

1 Origins of Humanity

How did the human species arrive at its present form? And how does our species history bear on the nature of this form? The human species, as it now exists, possesses language, rational thought, culture, and a specific affective makeup: but there was a time when our ancestors had none of these things—or had them only to a very limited degree. How did we get from there to here? How did we become what we so distinctively are, given our early origins? To put it more philosophically: how is *Homo sapiens* possible?

This question belongs to the field of paleoanthropology, and a rich body of knowledge is associated with it, mainly based on fossils.[1] More broadly, it is a question of evolutionary theory. It has not much attracted the attention of philosophers. This is a book about philosophical paleoanthropology. It attempts to link questions from an empirical science to more strictly philosophical concerns. So it is interdisciplinary. The disciplines linked are anatomy, evolutionary biology, archaeology, linguistics, psychology, cultural studies, and philosophy. I would describe it as a work of "evolutionary philosophy" (*evo-philosophy*, for short): in it I philosophize about the science of evolution. If we wish to have a label for this composite field, analogous to the recently minted "cognitive science," we might do worse than choose "emergence science." We are trying to understand how a certain suite of characteristics emerged in a particular primate species, using all available resources. No other species has evolved these characteristics, at least to the spectacular extent that we have, and it is a puzzle how we *could* have evolved them. What might explain the remarkable emergence of humanity?

The book therefore deals with an explanatory question: how can we explain the transition from our early apelike ancestors to human beings as we are today? This is a big gap to bridge—an explanatory gap. Much of the discussion will concern what *sort* of explanation is acceptable—that is, with conditions of adequacy. I will lay down various explanatory metaprinciples. We are trying to reconstruct a piece of history—or prehistory—and we need to know what constraints our reconstruction must respect. Naturally, this will involve hypotheses about what did in fact happen in the distant past. But it will also involve inquiries into how things *could* have happened—with what philosophers call "rational reconstruction." How is it *possible* to get from one state of nature to another? What sorts of intermediate stages make sense (that is, do not violate well-motivated constraints on evolutionary explanation)? Needless to say, the investigation is highly speculative, given the nature of the case. The form of the explanation consists in identifying and articulating the primary adaptations that powered the transition from early man to contemporary man (for short I will call this "the Transition").[2] This is simply the question of how contemporary humans evolved from ancestral stock that had very different characteristics (though also clear similarities). How did humans become the special kind of ape that we are? What accounts for the *difference* between us and other extant apes? Despite the considerable attention that has been devoted to that question, it is still not well understood.

The crucial role of the *hand* in producing and constituting human nature has long been recognized. Charles Bell published his classic text *The Hand: Its Mechanism and Vital Endowments as Evincing Design* in 1833, which argues that only a divine creator could have made something as wonderful and fitting as the human hand.[3] In *The Descent of Man* (1871) Charles Darwin drew a different lesson from the marvels of the hand: "Man alone has become a biped; and we can, I think, partly see how he has come to assume his erect attitude, which forms one of the most conspicuous differences between him and his nearest allies. Man could not have attained his present dominant position in the world without the use of his hands, which are so admirably adapted to act in obedience to his will" (84). For Darwin, the anatomical adaptation that is the human hand is the chief engine of human emergence—it is what made us the remarkable creatures we are.[4] The hand is the source of our biological success, our species ascendancy. This position was later developed in greater detail by John Napier, the physician and primatologist, in his books *Hands* (1980) and *The Roots of Mankind* (1970) and other works, drawing upon evidence unavailable in Darwin's time. It is now something of a commonplace in paleoanthropological studies.[5] Frank R. Wilson, in *The Hand: How Its Use Shapes the Brain, Language, and Human Culture* (1998), writes as follows: "It is genuinely startling to read Bell's *Hand* now, because its singular message—that no serious account of human life can ignore the central importance of the human hand—remains as trenchant as when it was first published. This message deserves vigorous renewal as an admonition to cognitive science. Indeed, I would go further: I would argue that any theory of human intelligence which ignores the interdependence of hand and brain function, the historic origins of that relationship, or the impact of that history on developmental dynamics in modern humans, is grossly misleading and sterile." In the present book, I too will sing the praises of the hand, though in a more philosophical key, finding an essential explanatory role for it in human evolution, particularly with respect to language. The central thesis of the book can then be very simply stated: it is the human hand that accounts for the Transition. It will turn out to be a little more complicated than that, to be sure, but the basic idea is that the hand made us what we are (indeed, *that* we are). The present author, like those preceding him, confesses to being enamored of the human hand, the often-neglected core of our humanity. The hand is what may have saved us from early extinction and what became the foundation of all our subsequent achievements. We owe it everything, quite literally. That, at any rate, is the general thesis, to be elaborated (and qualified) as we proceed.

I decided to call the book *Prehension* (not *The Intelligent Hand* or *Hand and Mind* or some such) for two reasons. One is that I wanted to fasten onto a particular function of the hand, namely gripping or grasping, which I regard as especially important. The other is that the hand is not the only prehensive organ in nature and I wanted to consider the role of prehension in evolution more generally (the mouth and the mind, in particular).[6] The hand for humans is the dominant prehensive organ, but prehension itself is a much broader phenomenon. Still, the hand will play an absolutely central theoretical role in what follows, since we are concerned mainly with the human species.

I will not enter into elaborate justifications for the claims made by paleoanthropologists, nor worry too much about details, but simply accept the broad outlines of what they say. My main interests concern the interpretation and ramifications of the established science (insofar as this science can ever be regarded as established). I do, however, recommend some immersion in the scientific literature for my philosophical readers, if only to accustom them to the style of the field.[7] I am interested in the philosophical consequences and uses of their findings and theories. I seek to bring their *mode of thinking* into the philosophical mainstream.[8] This means that I fully accept the standard account of human evolution, even though it is incomplete in many ways. I will, however, allow myself much more freedom in describing early human life than scientists typically permit themselves—especially with regard to the psychological condition of early man. It won't be all fossils and external behavior, but will also discuss what it must have *felt* like to undergo the kinds of transitions that we know occurred. I am concerned with the psychological evolution of man—the evolution of man's soul, if you like—as well as with his anatomical and behavioral evolution.

Heuristically, we can picture the inquiry as retracing our ancestry back to prehuman times. Take your parents and then their parents and then their parents, and so on going back through hundreds and thousands of generations. As we travel further back in time, there is a more and more pronounced divergence between you and your ancestors: where they live, what size they are, their expected life span, their coloring, their technological condition, their social arrangements, their use of language, their mental sophistication. If we go back far enough, now finding ourselves on the African continent, some millions of years ago, we will find our apelike ancestors, occupying the copious tropical forests and looking and behaving a good deal like modern apes. What we want to know is how these successive

generations—these sons and daughters of fathers and mothers—
became modified over time to become the human being that you
now are. We want to describe the contours and driving force of
the Transition. What were the selective pressures, the resultant
adaptations, the cognitive and affective modifications? In this
long ancestral sequence we will keep our eye firmly on the hands
of the successive generations—on their anatomy and activities.[9]

2 Two Evolutionary Principles

As a preliminary to our later discussions, I shall now enunciate two basic principles governing the evolutionary process. The principles are closely related and are commonplaces in evolutionary theory. Given their familiarity, I do not think it is necessary to defend these principles, but clearly articulating them can be useful when considering various putative evolutionary explanations. I shall call these principles the *principle of ancestral preservation* and the *principle of incremental adaptation*.

The first principle is well captured in a resonant and carefully honed phrase used by Darwin in the very first sentence of *The Descent of Man*: "modified descendant of some pre-existing form." Darwin is about to argue that man is just such a modified descendant of a pre-existing form, namely an ape, and he does so by observing that man shares many of the characteristics of (other) apes. The hypothesis is that this is because the process of natural selection has *preserved* the earlier forms in the later forms. Natural selection has operated on an ancestral stock and modified that stock in various ways, but the characteristics of the earlier stock have persisted though time and now (partly) constitute the present stock. In other words, preexisting forms have been *inherited*, so that there are "remnants" of the earlier creatures in the later ones. These remnants may be adaptive or neutral or even slightly disadvantageous; but the reason they are there is simply that the earlier forms had them and passed them on genetically. Later forms modify earlier forms, retaining them over time.

This principle stands in contrast to the theory Darwin is opposing, namely that each species was created separately. Here there is no preservation of ancestral traits, simply because the present species did not evolve *from* an earlier species. There is no mechanism whereby the evolutionary process must use earlier forms as raw material on which to build later forms, since all species are created independently. Thus, according to the "separate creation" theory, it is just an accident that species share many essential characteristics (God could have created them as all completely unlike each other). On Darwin's theory, by contrast, the present embeds the past, so that later species retain and contain features of earlier species. The long evolutionary history of a species is recorded within the present makeup of the species. The more recent the descent, the more will be preserved, but even remote ancestors can leave their imprint on subsequent creatures, marking them as having had a particular evolutionary history. This preservation of features will be most evident in anatomy, but in principle it also applies to behavior and psychology. Traces of the past will persist in the later creatures, shaping their bodies, minds, and behavior. Thus biologists often speak of "vestigial" features in human organisms that no longer serve any function—the appendix, the earlobes, body hair, toe nails perhaps. But even the most adaptive organ is an inheritance from ancient forms; it is not created anew when a species evolves. For example, the feet and wings of mammals and birds ultimately derive from the fins of fish, and accordingly share much of their anatomy with earlier fins. Apparently, the bones of the inner ear in mammals derive from the jawbones of ancestral reptiles.[1] Quite literally, earlier forms, often of animals now extinct, are alive in present species. Evolution works with the materials it has, and its mechanism is the modification of these materials over time. Every species now in existence derives from a preexisting species that has been modified over time; no species is created *de novo*, and no organ lacks its evolutionary lineage. The evolutionary process brings in the new by tinkering with the old, and the old is preserved—pickled, we might say—in the new.

This preservationist picture can be compared with the growth of a city over time. There is the most ancient part of the city, often crude and simple, but this has been worked over to create more recent structures—not demolished and replaced, but modified. A single building might thus incorporate architectural forms from widely separated eras. What was once a stable might now be converted to housing; a jail might become a hotel. Relics from the distant past are everywhere, mixed in with more recent structures. The old persists within the new, sometimes usefully, sometimes accidentally. The difference between a city and a species is that the city can be intentionally planned, but evolution works by trial and error. And evolution *can't* demolish a species and then replace it with another from scratch (as God could if He didn't like the first species He created). Evolutionary change is thus inherently conservative and haphazard. It is rather like geological change: one stratum of rock can be superimposed on another, but the original rock survives, sometimes hidden, forming "bedrock." The deeper you dig, the further into the past you travel. This is why it is often noted that human DNA contains the remnants of DNA from ancestral species, which may no longer be active, but are preserved because they are not harmful to the organism. There is no reason why this principle should not be as true for the mind as for the body—relics of earlier minds may persist in descendant minds, even if no longer accompanied by the original environment in which they evolved. There is a kind of "law of inertia" at work in the evolutionary process—persistence unless actively thwarted.[2]

The principle of incremental adaptation states that evolution works gradually, not radically. There are no sudden leaps forward, no abrupt revolutions. In traditional evolutionary language, there are no "saltations"—there are only "incrementations" (to introduce a neologism). This principle too is implicitly contained in Darwin's pithy formulation, "modified descendants of pre-existing forms"—evolutionary change consists of modifications, not outright inventions. The textbook example is the giraffe's neck: it did not change from short to long in a single evolutionary event, but gradually lengthened over a considerable period of time, as small mutations were positively selected for, one after the other. The lengthening was cumulative, not abrupt and unprecedented. Much the same can be said about an elongation more germane to the subject of this book: the human thumb is a lengthened version of earlier primate thumbs, and the lengthening was gradual, not sudden. It is easy to miss this point if you tacitly presuppose a "homuncular" view of evolution, according to which some sort of intelligence directs the process toward some chosen ideal end. But the process is not teleological in this way, so the end point is achieved "accidently," no matter how apposite it may appear. Each increment must confer an adaptive advantage in order to be passed on, and the sum of them confers a large adaptive advantage. It is logically possible that a single mutation might produce an unprecedented trait with a brand new adaptive advantage, but in practice the probability of this is vanishingly small. Thus a mouse might (as a matter of logical possibility) undergo some extraordinary genetic convulsion that results in forming a mouse brain as intelligent as the human brain, but such things simply never happen. Mutation is a very blunt and blind instrument and cannot be expected to work miracles: it offers up modest changes that natural selection either allows to persist or eliminates. We never get something from nothing in evolution. New species don't spring into existence overnight by dint of some spectacular mutation, and the same is true of new organs or traits. Evolution always takes the form of a gradient, not an upsurge of abrupt steps.[3]

Biologists have a useful term to express this fundamental point: *preadaptation*. Using this term, I can state the incremental principle in the form of a law: for every adaptation there must be a preadaptation. A preadaptation gives the *appearance* of anticipation, as if the lengthening neck or thumb were aiming at some ideal end-state of optimal length. But it is not real anticipation; it is entirely accidental, mechanical. The entire process is guided by blind natural selection, one piece at a time. It is just that every new adaptation needs a platform in the prior structure of the animal's ancestors—something to work with. If we want to explain how a given adaptive trait evolved, we have to identify the preadaptation from which it arose by small coherent steps—by incremental modification. It is never theoretically satisfactory to postulate a sudden leap in the dark that had a positive result; we need to identify the intermediate steps. Worms cannot evolve from bacteria in a single generation, and speakers cannot evolve from nonspeakers in a single generation. There has to be a bridge between the two states. This places a demanding constraint on evolutionary explanations: no saltations allowed! The evolutionary process works incrementally, so our explanations must respect this natural fact. That is basically why evolutionary change takes so long. It is thus a condition of adequacy on evolutionary explanations that they respect the principle of incremental adaptation. We need precursors, precedents, preconditions. Later we will see what this involves in particular cases, notably language; for now, I am just laying out some methodological desiderata. Very often it turns out that some prior trait that worked well enough in a given habitat now acquires a different adaptive function in a new habitat, for which it is then selected and perhaps modified. We shall see that this is conspicuously true of the hand, when the habitat of our ancestors changed quite dramatically.

The picture of evolution suggested by our two principles is then this: evolutionary change is tightly constrained by the past and proceeds by small adjustments going into the future. Earlier forms persist and are modified gradually over long periods of time. There are no sudden departures and revolutionary developments in real evolutionary time, and earlier forms set the parameters for later transformations (this is quite compatible with so-called *punctuated equilibrium*, by the way, once timescales are properly appreciated).

I can explain this picture by means of what I shall call *adaptive space*. Consider all logically possible adaptations to an environment consisting of sea, land, and air—an environment like that obtaining on Earth. This is probably an infinite set, or in any case an extremely large one. The set of actual adaptations on Earth is only a small subset of this enormous set. On the independent creation story, with God as creator, any of these logically possible adaptations could occur at any time, according to God's choice. There is no constraint stemming from the past or necessity for incremental modification. But our two principles tell us that it matters tremendously how the process begins. Suppose life begins in the sea and then proceeds to land and finally ascends to the air—fish, reptiles, and birds (to simplify). Given that fish come first, they establish the constraints on later developments, so that their form will be preserved in subsequent species of reptiles, mammals, and birds. For example, the pectoral and pelvic fins of fish will act as the templates for reptilian and mammalian feet and for avian wings. Later species will accordingly be tetrapods, with each limb ending in extremities possessing a five-pronged ray structure. This will be the basic animal design, preserved across newly evolved species. But if we imagine a planet on which life began in the air, say, because of special conditions obtaining there, and then moved down to the land and finally to the sea, then we will find a very different evolutionary trajectory, in which birdlike features are preserved down through the generations. The fins of fish will resemble the wings of their early bird ancestors, not vice versa.

Adaptive space will take on a different shape on this planet from that obtaining on Earth. The preexisting forms will be quite different, and the modifications will preserve them to some marked degree. If we suppose that finally a species evolves that possesses language, reason, a rich social life, art, and science, this species nevertheless will not resemble us very closely (except in those respects)—since it will preserve different preexisting forms. The basic body plan might incorporate eight limbs with twelve fingers on each of four hands, with vestigial feathers and a high chirpy voice. Any given evolutionary setup will trace a particular path through total logically possible adaptive space, and this path will reflect the starting point. It will not proceed according to what would be ideally adaptive at any given time, as we might expect on the creationist model, but according to the available existing materials for mutation and natural selection. There will be fixed initial conditions and then evolutionary trajectories that preclude saltation. The fact that a particular species will go extinct at a certain point unless it acquires trait T has no tendency to make that species acquire T—however "nice" that would be. There is no benevolent supervisor of the evolutionary process; there is just small-scale random mutation and (non-random) natural selection working on mutational input. The trait T may be highly adaptive, even indispensable to survival, but it is not an "accessible trait" for that species at that time. At some later time, after much incremental variation, T might *become* accessible; but it is not accessible unless sufficient *pre*adaptations are already in place to provide a platform for its development.

Thus, high intelligence might be a generally adaptive trait for many species in many environments, but it is not an accessible trait for the vast majority of species, simply because there is too much discontinuity between the present form of the species and that desirable trait. To explain how a trait arises (say, language), we need to show that it is accessible given the antecedent attributes of the organism—that is, we need to tell an incremental story.[4] We cannot just announce that a very fortunate adaptation occurred in which the entire apparatus of human language was suddenly and spectacularly installed by freak mutation in the human brain. If we are trying to give an evolutionary explanation of the Transition, we need to respect the principle of incremental adaptation and recognize that antecedent forms constrain present potential. It must be a matter of small steps, fine adjustments, and steady progress, with previous traits constraining the entire process—smooth ascent, not sudden elevation. As I remarked at the beginning of this chapter, these points are quite orthodox in evolutionary thinking, but it is worth making them maximally explicit, especially when we come to consider the evolution of cognitive traits, such as language, that appear to have no obvious antecedents in the rest of nature.

3   Human Prehistory

In a famous passage from *The Descent of Man*, Darwin writes as follows:

As soon as some ancient member in the great series of the Primates came, owing to a change in its manner of procuring subsistence, or to a change in the conditions of its native country, to live somewhat less on trees and more on the ground, its manner of progression would have been modified; and in this case it would have had to become either more strictly quadrupedal or bipedal. Baboons frequent hilly and rocky districts, and only from necessity climb up high trees; and they have acquired almost the gait of a dog. Man alone has become a biped; and we can, I think, partly see how he has come to assume his erect attitude, which forms one of the most conspicuous differences between him and his nearest allies. Man could not have attained his present dominant position in the world without the use of his hands which are so admirably adapted to act in obedience to his will. As Sir C. Bell insists "the hand supplies all instruments, and by its correspondence with the intellect gives him universal dominion." But the hands and arms could hardly have become perfect enough to have manufactured weapons, or to have hurled stones and spears with a true aim, as long as they were habitually used for locomotion and for supporting the whole weight of the body, or as long as they were especially well adapted, as previously remarked, for climbing trees. Such rough treatment would also have blunted the sense of touch, on which their delicate use largely depends. From these causes alone it would have been an advantage to man to have become a biped; but for many actions it is almost necessary that both arms and the whole upper part of the body should be free; and he must for this end stand firmly on his feet. To gain this great advantage, the feet have been rendered flat, and the great toe peculiarly modified, though this has entailed the loss of the power of prehension. It accords with the principle of the division of physiological labour, which prevails throughout the animal kingdom, that as the hands became perfected for prehension, the feet should have become perfected for support and locomotion. With some savages, however, the foot has not altogether lost its prehensile power, as shewn by their manner of climbing trees and of using them in other ways. (84)

The basic outline of human prehistory, outlined in this pregnant passage, has not much altered since Darwin's day, though he was unaware of the many hominid species that have subsequently appeared in the fossil record. I can summarize current thinking roughly as follows. Our primate ancestors were at one point primarily tree dwellers. At this stage there was probably little in the way of language or social cooperation, and the body would be adapted for tree climbing, branch suspension, and brachiation (swinging from branch to branch). Food would likely consist of fruits, nuts, and insects. All limbs would be prehensile, with little to differentiate the feet and hands. Tool use would be virtually nonexistent.[1] Trees would afford shelter from sun and rain, as well as safety from ground-dwelling predators. This lifestyle would have been in place for millions of years, stable and satisfactory—just as it still is for many primates. Modern humans are descended from this ancient stock of arboreal ape like creatures, subject to the principles of ancestral preservation and incremental adaptation. Simplifying, we can say that "we" were once tree dwellers, though at this stage our ancestors were not yet real *Homo sapiens*.[2]

But then, for some reason, we came down from the trees (our ancestors did): maybe climate change affected the food supply in the trees, maybe the babies became too big and heavy to carry in trees, maybe there was competition from other species, or maybe it was a combination of things. In any case, we (they) descended to the ground, possibly over many thousands of years, maybe more quickly. Our habitat thus changed dramatically. At first we had only the old adaptations to arboreal life, which now had to serve us in terrestrial living. This initiated a long process of natural selection, probably fairly brutal, in which mutations that favored terrestrial living were powerfully selected for. Darwin sketches the general form that these took: the bipedal posture, straight-legged striding, differentiation of hands and feet, development of the hand as manipulating organ, different modes of feeding, child care, and so on. What we have here is a major change of habitat interacting with an old species design, which produced intense selective pressure. Our bodies (and brains) are terrestrial modifications of a preexisting arboreal form.

After many thousands of years (estimates vary), man became a dedicated toolmaker, a language user, and a sophisticated social being—and then later an artist, a scientist, a philosopher, and do on. Above all, he developed a large and complex brain (*encephalization*). This is what I am calling the Transition. The theory suggested by Darwin is that the hand played a critical role in making the Transition possible—it was the bridge, the fulcrum, the mediating mechanism. The hand was once an organ for locomotion, branch gripping, food gathering, and probably fighting, but it grew into another kind of organ altogether. These later adaptations of the hand (of which more later) were built upon the preadaptations present in its arboreal incarnation—gripping fingers, basically. The hand became liberated from locomotion by upright bipedal posture, and it came into its own to serve other adaptive functions.[3] This led to a process of coevolution between hand and brain, as the brain increased in capacity to serve the hand and the hand developed new powers because of the expanding brain. The disadvantages of the bipedal gait—problems of balance, spine fragility, slow running—were eventually compensated for by the adaptiveness of the new arrangement. Free hands were far more vital to survival on the ground than the disadvantage of easily tripping over on just two feet (and our brains had to become better at keeping us thus balanced, which led to further enhancements).

Thus, in sum, we became what we are today because we were driven down from the trees and forced to adapt to a new and hostile world. If life in the trees had remained hospitable, we would never have made the Transition, as our primate relatives have not. Our hands would not have been transformed into the prehensile prodigies they are today, had we retained our ancient habitat. What looked like a catastrophe turned out to result, in the fullness of time, in an unprecedented species success, largely because of what the hands became under the new selective pressure. We snatched victory from the jaws of defeat. What looked like impending extinction eventuated in unparalleled species proliferation. And the hands played a pivotal role in this reversal of fortune. They allowed us to claw our way back after a shocking and perilous deracination.[4]

That is the general picture I will be working with. It will be spelled out more fully as we proceed, and its implications explored, but the basic message should be clear: man is an ex-brachiator, an expelled tree dweller, a hand specialist, a recent (and unlikely) biological success, a touch-and-go proposition for much of his early existence, and a creature still very much in the making.[5] The question is how to understand man's later evolution given his inauspicious beginnings, both as to form and fortune. How, in particular, did the hand contribute to the development of language?

4   Characteristics of the Human Hand

We are deeply familiar with our hands, though we may not pay them much conscious attention. We take their remarkable powers for granted. In particular, the relationship between the thumb and the fingers is critical to their function. John Napier writes: "One cannot emphasize enough the importance of finger-thumb opposition for human emergence from a relatively undistinguished primate background. Through natural selection, it promoted the adoption of the upright posture and bipedal walking, tool-using and tool-making that, in turn, led to enlargement of the brain through a positive feedback mechanism. In this sense it was probably the single most crucial adaptation in our evolutionary history" (*Hands*, 55). In the human hand, by contrast with the hands of other primates, the thumb is much elongated and the fingers correspondingly shortened, so that the thumb can oppose all four fingers. The thumb is also much more mobile and muscled in humans. Being able to hold and squeeze various objects between the tip of the thumb and the tip of the index finger is unique to humans. The thumb of a chimpanzee is a mere runt compared to the large powerful thumb of a human. With this hypertrophy of the thumb comes the potential for all sorts of grips unavailable to other primates. Yet this human hand evolved by slow modification from the primate hand—it is not a radically novel structure by any means.[1] Selective pressures must have encouraged its growth, power, and mobility. The big meaty thumb is the lever of our biological ascent, as Napier remarks.

The hand had its distant precursor in the lobe fish, whose fins have a pentadactylic structure.[2] This is a clear instance of ancestral preservation. But, as every anatomist delights to note, the human hand is an extraordinary complex of bones, muscles, and nerves, as well as pads and tiny ridges. It is clearly highly adapted, down to the last intricate detail. This complexity gives rise to great versatility, and the hand is capable of a large variety of grips. Napier distinguished the "power grip" from the "precision grip": the former is used to grip a hammer or tennis racket and involves the fingers, thumb, and palm; the latter takes place between the pads of the thumb and fingers, as when holding a pen or a strand of hair. The precision grip distinguishes human hands more than the power grip. Much human tool use involves the precision grip, which gives the hand a more delicate range of manipulations. Presumably this type of grip evolved later than the more brutish power grip. Writing is made possible by the precision grip, as is rolling small objects between the fingers, not to mention surgery. Of course, it is possible to alternate power grip and precision grip and also to combine elements of both in a single act of manual prehension. The anatomical complexity and functional range of the human hand, exquisitely combining power and precision, give it enormous versatility and control. No artificial hand is anywhere near matching what every human hand can achieve without its owner even thinking about it.[3]

Not surprisingly, a large quantity of brain tissue is given over to the hand: the organ clearly needs a powerful computer to supervise its operations. In a cortical homunculus diagram, the thumb by itself is typically depicted as about the size of an entire leg: that is, the same amount of cortex is devoted to the thumb as to the leg. But that is entirely predictable given the wide range of activities in which the thumb is engaged. Some of this cortical machinery is dedicated to motor function and some to sensory function. Not only is the hand a motor champion; it is also highly innervated for sensation and perception. And the two are intimately connected, as sensation in the hand guides the hand's movements. But that isn't all: in addition to its motor and sensory prowess, the hand is highly *educable*—so it must possess a powerful memory. The hand can be trained to remember astonishing sequences of movements—as with playing the piano or other instruments.[4] Much of human civilization depends upon the educability of the hand, which is to say on hand memory. This capacity must have evolved too, probably to aid in sophisticated tool use. No other primate exhibits this degree of hand know-how, as well as hand plasticity. The hand learns complicated lessons from experience. It is really not possible to isolate the hand as a functional organ from the brain that controls it—we may as well reckon the brain as part of the hand (the hand system: compare the visual system). The brain is the engine that drives the wheels.

The abilities of the hand are not all acquired, however. There must surely be a strong innate component. Hand competence is essential to human survival and hence is best not left to chance. The human infant develops the basic grips and skills without instruction, so that an innate program must be controlling the proceedings. It is rather like language: innate universals specialized into specific forms. You are not born with the ability to play the piano or tennis—these require specific training—but the basic component hand skills have an innate foundation. There is, we may assume, an innate prehension program—a universal "grammar" of grips and manipulations. A human no more learns how to squeeze an object between thumb and forefinger than a fish learns how to swim (which is itself a kind of gripping of the water). The same must be true of other primates and their hands (gibbons must have a "brachiation gene"), but human hands have a far richer innate program, to go with their more elaborate structure and function. In fact, the human hand's capacity seems to exceed the capacity of the ape hand to about the same degree that human language exceeds ape language (we shall see that these are not unconnected facts). The human hand bears a strong physical resemblance to the ape or monkey hand—not surprisingly, given the evolutionary lineage—but if you look beneath the surface whole continents distinguish them. The human hand is in another league, in terms of motor capacity, perceptual sensitivity, educability, and cortical dedication.[5]

It is important not to limit the hand to the anatomical structure on the distal side of the wrist. The wrist itself must be regarded as functionally part of the hand, and the forearm, with its tendons and muscles reaching to the fingers, is also a part of the hand's operation. Indeed, the entire arm and shoulder are typically involved—the hand does not act in isolation. We can even regard large segments of the human body as geared to the actions of the hands, from the feet up. The body is in many ways the hand's platform—its base of operations. So the body of man has adapted wholesale to the manual lifestyle, enabling the hands to go about their vital business more efficiently. When the hands came into their own, far exceeding their traditional occupations, some time after our descent from the trees, the body had to go along for the ride; the better the body served the hands, the more it would be selected for. Even the eyes must have felt this selective pressure, because so much of the work of the hands needs the eyes to keep it on track. You can only seize an object accurately if your eyes can first locate it. The human body as a totality is designed for supporting manual prehension, in a kind of anatomical holism. Hand and body are fully integrated. There is an extensive hand *system*. It is almost as if large segments of the body, from feet to trunk to shoulders to wrist, are components of the hand. It is convenient anatomically to separate the hand from the rest, but functionally the hand extends much further. Just consider the act of throwing: the missile is gripped between the fingers but the entire body goes into the act of throwing—you throw with your feet as well as with your hand. The body, we might say, is the handmaiden of the hand—its infrastructure.[6]

Clearly, the hand of our ancestors was subjected to strong selective pressure after our fall from the branches, or else it would not differ so significantly today from the hands of other primates. It was evidently able to change its structure and function as a result of environmental pressure. We can surmise that this is because the hand was so critical to survival, so that the pressure to modify was particularly intense. What an early human did with his or her hands made the difference between life and death, reproduction and its lack, so that the hand could not afford to remain limply unchanged. The hand became the *focus* of natural selection during those precarious millennia. Other human organs might not have been under such intense pressure—the eyes and ears of humans do not seem markedly superior to those of other apes—but the hands had to adapt more rapidly and dramatically. The lengthening and strengthening of the thumb must have been especially urgent. Fortunately, the hand was able to absorb the selective pressure and emerge a much more impressive organ than before. We went from hand mediocrity to hand genius in fairly short order, by the standards of evolutionary time. No doubt, we now have the best pair of hands on the entire planet.[7]

That, in crude outline, is the science of the hand; but what of the philosophy? How, to begin, should we *define* the hand? This question does not prove easy to answer. It obviously won't do to define the hand as "an organ with four fingers and a thumb," because there is no conceptual necessity to having that many digits—there are logically possible hands that have a different number of fingers and an extra thumb, or no thumb. Also, it would be nice to have a definition of "finger" and "thumb." We might try going functional: a hand is an organ used for gripping. That looks like a necessary condition, but it is clearly not sufficient: you can grip things in your mouth or between your knees. Then let us say that a hand is an organ that grips by means of fingers, which mouths and knees don't. But what is a finger but a part of a hand? To avoid circularity we need to define "finger." Now things get tricky: is the notion *finger* primitive? We can't say a finger is that which points, because not all fingers do and some non-fingers do (you can point with your elbow). Nor can we say a finger is part of a gripping mechanism, because lips and teeth are also part of a gripping mechanism. Perhaps this will serve: a finger is a flexible jointed protrusion that enables gripping. We thus rule out lips and knees as fingers. Are the octopus's tentacles fingers then, according to this definition? We might want to add that the protrusions need to be affixed to a palm, defined as a flat slab or some such, and that they need to be jointed. Or we might allow that tentacles *are* fingers. Are toes also fingers—of the foot? What if you are a foot acrobat using your toes to perform the functions of fingers? Are not these flexible toes "the fingers of the feet"? What we have captured in our attempted definition is a certain morphology linked to a function, and that is the heart of the notion of a hand, even if some cases are borderline or debatable. Martian extremities with a hundred prehensile protrusions would count as hands by this definition. Are the wings of a bat also hands? What if they are used for gripping? There can surely be webbed hands and feathery hands and slimy hands (the fins of some fish appear quite handlike). Similar problems surround the definition of other organs of the body—the mouth, the ears, the leg—because ingenious counterexamples can always be generated and intuitions may waver. But we need not be very concerned about these definitional problems here: our concern is primarily with the human hand, as it now exists, not with all logically possible hands (though this is a perfectly interesting conceptual question). Let us content ourselves with saying that the hand is a kind of morphological-*cum*-functional entity: a prehensive array of protrusions, roughly. But with that remark, I will leave the conceptual analysis of the hand to one side in this book. We certainly have a solid enough conception of what a hand is in order to proceed with our inquiries. And we are concerned with the role of the hand in human evolution and human life, not with all conceivable hands for all conceivable handed beings. So let us focus on the actual properties of the human hand. Our question, then, is what gave rise to them and what they in turn gave rise to. What forces shaped the human hand, and what did this hand enable?

5 Hands and Tools

What did the freed hands do for us once we descended from the trees? We no longer needed them full-time for climbing, clinging, and brachiating, so we could use them for other purposes—but what? Carrying babies, making fists, fighting, and scratching itches: but this they were presumably doing already. By themselves they offered little radical advance, and merely dangled. Tool use is the answer. It may be that some tool use already existed in the trees, as it now exists in other primates, but the stage was now set for a great expansion of this activity. Hands turned seriously to tools. Gripping branches would naturally lead to primitive tools, as branches would break and be left in the animal's grip, ready to be used as clubs, say. Even unbroken, the branch is functioning as a tool for the animal—an instrument whereby the animal achieves its goals. In any case, as anthropologists have demonstrated, the hands came to be used extensively in the manufacture and employment of tools.[1] The tool replaced the tree as man's chief object of prehension: he went from gripping one kind of thing to gripping another, both in the service of survival. Or, as we might put it, man *resorted* to gripping tools, given that his old arboreal habitat was no longer available. His hands were bereft and his survival correspondingly threatened; thus he needed to find other work for these now-idle hands, and to discover new survival tactics.[2] Thus began the era of man the tool-user (no longer the tree-user). And the hands were the part of the human body with which tools immediately interacted. The "hand-tool nexus" was formed and consolidated.

A huge amount has been discovered about early human tool use, but I will not be concerned to rehearse any of this, interesting though it is; I shall limit myself to some interpretative philosophical remarks. First, how shall we define the notion of a tool? If we define a tool as any part of the environment used by an animal to aid survival, we include far too much: the entire habitat of the animal will count as a tool—trees on which birds perch, food materials, water swum through, air breathed. The Sun will count as a tool, as will gravity and the ground walked on. The *OED* defines "tool" as "a device or implement, typically hand-held, used to carry out a particular function," or "a thing used to help perform a job." The second definition is extremely broad, unless we interpret "job" very narrowly (as in "paid occupation"). The first definition is also very broad, since many objects can be "used to carry out a particular function"; so we need to build something more restrictive into the words "device" or "implement"—we have to take these words to be limited precisely to *tools*. Isn't a twig a device for perching on, as far as a bird is concerned? Isn't a branch an implement for a brachiator to use to move around? A branch is certainly hand-held by a gibbon, say. It is not clear whether there is any useful natural kind here. Also, we should not adopt a behaviorist conception of tool use, on which a tool-user is simply a creature that externally operates what is in fact a tool. If a bird perches on a hammer, it is not thereby a tool-user. Nor should we require that the animal *make* the tool. This is not a necessary condition, because some tools are found, not made. Nor is it sufficient to identify what is special about human tool use. Ants make ant nests, but is this the notion of tool that we need when considering human tool use?

When we consider the human construction of axes and spears and the like, and their subsequent employment, what is it that impresses us in this behavior? It is not merely that these are objects with functions made by man—birds' nests have functions and are made by birds. What impresses us, surely, are not the material facts by themselves but the *cognitive background*. It is that a certain kind of *thinking* lies behind the observable behavior—a certain mode of conceiving the world, and hence altering it. It is not the external behavior of using tools that sets us apart, still less the material form of the tool, but the psychological processes that drive this behavior. It is our tool-using *intelligence*. Even if these processes never actually eventuated in practical tool using, they would still set us apart from other animals. There is all the difference in the world between operating an implement with this cognitive background and operating an implement without it. My point here is not to deny that other apes genuinely use tools; it is just to say that what is important, in terms of evolutionary advance, is whether they share our cognitive processes. They may share them in rudimentary form, in which case they have the beginnings of what is crucial—the external behavior is not the point. Let me put it this way: a mindless zombie cannot really be a tool-user in any interesting sense, even though externally its behavior may look a lot like ours. To have what early humans had when they made and used tools you have to have a certain kind of mind.[3]

What, then, is this special cognitive background? I suggest that it has two aspects: creativity and teleological reasoning. Thus a real tool-user, such as early man assuredly was, must be a being capable of creative thought and teleological thought. Birds and ants may build impressive structures that function to aid their survival, but they do not engage in creative teleological thought. *This* is what makes tool-using early man stand out from such other species. This is what powers his progress toward his final destination. It is not the tools as such but the tool-using state of mind—the internal, not the external. By "creativity" I simply mean the ability to see a new use for a piece of found material or the ability to construct something having a new use. It is not enough to instinctively use a piece of the environment for a certain end—say, a twig as part of a nest. What distinguishes the early human use of tools is *ingenuity*, the seeing of fresh possibilities, imagining what can be done with an object.[4] The analogue for birds would be, for example, suddenly stealing human hats to be made into nests, as a result of observing the properties of hats and reasoning that they would make good nests. There has to be an element of invention—not in external tool use as such, but in the kind of "tool cognition" present in early man. There has to be the idea of a creative solution to a felt problem—as in attaching a handle to an axe head to increase percussive power. The mere behavioral use of tools is not the issue; it is what went into producing and using the tool that counts. There has to be what we call "intelligence"—creative problem solving, use of the imagination—not the mere manipulation of foreign objects. If an animal finds a new use for its claws, never envisaged before, showing real ingenuity, then that is the relevant talent. The use of stems by apes as "fishing rods" to extract termites from their nests is the right kind of thing to qualify—a genuinely novel solution to a problem. This is what we can assume early humans to possess, though humans now greatly outclass apes as tool-using animals.[5]

The second aspect, which is connected to the first, is that the animal must be able to engage in systematic means-end reasoning. There is a goal to be achieved and a means is devised to achieve it. This requires having teleological concepts: *end, means, working toward an end*. A future desirable state is envisaged as such and present action is intentionally directed to achieving that future state. This will typically involve deferred gratification: the thought that the future will be better if we make tools now, even if there are more fun things to do at present. This is a complex and sophisticated mode of reasoning. Thus the creativity must be woven into teleological rationality: being creative about the means to achieve a desired goal. The agent must have the *idea* of a tool—know what a tool is—and must also grasp the concept of work: she must *understand* that making *tools* requires *work* for achieving future *goals*. This in turn requires a degree of self-consciousness—thinking about what will be good for *me* (or us). The agent must grasp the nature of planned action, calculated decision. She must also see the world as a potential repository of tools—that conception must be applied to the world of natural objects. She must have thoughts like: "I can make a tool of this." She doesn't just use tools; she *thinks* tools. She has tool intentionality, tool consciousness, tool being-in-the-world. Her inner psychological being is that of a toolmaker and user. This is the mode of cognition that early humans so clearly display in their construction of tools, and which some apes shows glimmers of today. It is not merely that they use bits of the environment to perform certain functions: that is common in the biological world, and is not particularly impressive or distinctive. But early man *thinks* in terms of instrumental functions: about what is useful, how to achieve it, the actions to be performed, and the sacrifices that need to be made. He views the

world instrumentally, and he is ready and able to employ ingenuity to make better instruments. Tool conceptualization is the
important adaptive trait, not merely tool behavior.[6]

Once the tool-using adaptation is in place in rudimentary
form, the possibility for brain-tool coevolution emerges, as new
tools call for increased brainpower and increased brainpower
leads to more sophisticated tools. There will already have existed
a dawning tool consciousness in man, because in the trees
branches and twigs can be used as tools; but once ground living
became entrenched, the selective exigencies intensified. Humans
needed more and better tools, because their new environment
simply did not yield vital resources so readily as the trees. Tools
for hunting, gathering, and scavenging will have conferred an
adaptive advantage.[7] Containers for food carriage would be particularly useful. In addition, the freeing of the hands would have
led to an increasingly instrumental view of the hand itself—of
what it could be used *for*. New methods of using the hand were
called for by tool use, and hence increased manual ingenuity.
The hand may indeed have been the original conceptualized
tool, as early humans began to appreciate its instrumental powers—for instance, its utility as a device for carrying things over
long distances or as a digging implement for uncovering edible
roots. The fingernails themselves provide an excellent prototype for tools designed to score and tear—the ingenious early
human just needed to make sharper and more powerful nail-like
implements. Once the mind has become tool conscious and tool
seeking, everything becomes a potential tool, including parts
of the body (consider carrying things balanced on top of the
head). The human mind becomes tool saturated, tool obsessed.
The world comes to be viewed as one big toolbox. Anthropologists rightly stress the evolutionary importance of human tool

use, but I suspect that a lingering behaviorism has made them hesitant to ascribe a rich background of tool psychology to early man; they want to stay "objective." But, I would insist, what passes through the consciousness of ancient man—what occupies his daily thoughts and dominates his imagination—is as important evolutionarily as what passes physically through his hands. In understanding early humans, as well as our arboreal ape ancestors, we need to try to see the world as they did, which means reconstructing their conceptual scheme—their very mode of consciousness (we could call this *phenomenological paleoanthropology*). And what we find, as evidenced by their actual tool production, is a blooming tool mentality—the conceptual "instrumentalization" of the world, in effect.[8]

As the brain coevolved with tools, so the hands coevolved with tools. In fact, there was a three-way coevolution: hands, tools, and brains became reciprocally modified by positive feedback loops, as if trying to keep up with each other. A newly invented tool, say a spear, would need a particular design of hand and arm to be used in the most effective way, so that design would be henceforth selected for. But the modified limb needed a new control system in the brain, which led to brain enhancements that fed into further tool invention. New tools need new grips, but new grips lead to new tools, and so on. The human hand thus became ever more precise and versatile, as a variety of tools came into use. The great utility of tools in survival imposed a strong selective pressure on the hand and the brain to coevolve so as to make maximum use of these tools. The hand must *match* the tool it is gripping and wielding; there has to be an effective hand-object correspondence. The hand then slowly evolved over many thousands of years into a device for devices, a tool for tool use. It became fine-tuned by tools, made in their image.[9] Polymorphic manual prehension of tools was the key to human survival. Dexterity and versatility of the hands were at an adaptive premium.

With this gradual but momentous modification of the hand came another human modification: greater social cooperation. In the trees, food gathering could be more or less individualistic, but on the ground, where hunting and scavenging became the norm, organizing into groups had survival advantage. You could not expect to best a big cat or a mammoth on your own, or even a gazelle. Thus man had to become more social—he needed to belong to a pack. Tools were woven into this increased socialization: they could be exchanged, shared, allotted, portioned out; and the process of their manufacture would require both teaching and the division of labor. The productive role of the hand would be evident to all, and intraspecific hand competition would inevitably arise. The best-made weapon would go to the one with the most able pair of hands. The best toolmaker would be esteemed for his dexterous hands. The individual with whom to mate would be the one whose hands bring the most to the table. Thus hand pride would emerge, and hand admiration, and hand hierarchy. Social bonds would revolve around tool use and hand skills. At some point the hand was introduced prominently into sexual activity (other primates do not use the hand in this way, apparently, or not much anyway). Grooming with the hand would have its own natural appeal. The uses of the hand would thus range all the way from brute aggression (clubbing others) to tender grooming and stroking. Clearly, having powerful, clever, agile, sensitive hands would have social value, as well as individual value. It became the "survival of the handiest."[10] Thus we see a constellation of factors evolving in unison: tool use (with tool cognition), hand dexterity, brain enhancement, and social structures. This is quite a rich brew, entrained by our forced descent from the trees, where we could get by with less.

We now have a number of adaptations working furiously together in human evolution: bipedal locomotion, hand centeredness, brain expansion, social organization, tool creation and use, and an underlying cognitive sophistication. Each of these factors operates in conditions of dire necessity, because the descent from the trees was forced and extremely challenging. We had to adapt rapidly to the new habitat, and those ill suited to it would soon be extinguished. Presumably many perished in the years following the descent, so intense was the pressure. The population may have shrunk alarmingly.[11] The new adaptations were vital. None of these adaptations can be properly understood except in the context of the others: it is a total package. There is a kind of "holism" at work here. Man perforce became a big-brained, tool-using, hand-oriented, socially adept creative thinker. Tools were his answer to expulsion from the trees, where they had only a limited point, and tools brought many other changes with them. Where once he gripped branches he now gripped spears and axes (equipped with his spanking new enlarged thumb). His hands were the primary organs of adaptation to the challenging new habitat. Thus, we might say, our hands were born in the trees but were educated on the land. We had a preexisting arboreal manual form that became modified into a terrestrial form by incremental steps. In our hand we see the form of the ape's hand clearly inscribed, but we also see the cumulative effect of innumerable small changes—ancestral preservation plus gradual modification. This is all perfectly intelligible and nonmiraculous. What *would* be unintelligible would be the suggestion that a species quite devoid of hands would suddenly sprout them to cope with an environmental need. Such a saltationist story is, of course, quite preposterous, and we have no need for it: there is a clear continuity between the ancestral form during our arboreal past and the terrestrial human form that exists today. The tree-climber and brachiator is already using his hands by gripping branches; his descendant employs the same basic hand capacity when grasping a spear. A slight elongation of the thumb, combined with some muscle enhancement at its base, will then enable a wide range of new prehensive skills. No saltation is being contemplated here.

As Darwin says, the liberation of the hands entailed by the bipedal gait is the decisive event in our transition to dedicated tool-users. A quadrupedal gait would not have been consistent with the ascent to a full-time tool-using animal. The shrinkage of forelimbs into arms that dangle from an upright trunk led to a huge expansion of the brain, because the small protrusions called "fingers" became vital to our success as a species. The human arm is not an atrophied leg, as with some bipedal dinosaurs and contemporary kangaroos, but a strategically streamlined adaptation. We have the arm that our hands need. It is the redeployment of the hand in the direction of tools, coupled with hand-driven encephalization, accompanied by hand-mediated socialization, which accounts for the Transition, according to the present Darwin-based theory.

But how does this account for the origins of language? Nothing we have said so far has any obvious bearing on that question. Language emanates from the mouth, does it not, so how can the development of the hands be relevant to that? How do we make the transition from manual tool using, however sophisticated, to vocal language using? How do we get from hands to speech?[12] That is the topic of the next chapter.

6 Hands and Language

Language as it now exists has obviously evolved over many thousands of years. We cannot expect to provide an intelligible simple bridge from prelinguistic man to language in its present form. What we must look for is a theory of the origins of the basic form of language, from which it may evolve into what we see today, suitably supplemented. I am concerned here only with this initial evolution. But even here saltation threatens: how do the basic syntactic and semantic structures of language originate? What is the preexisting form from which language arises by incremental modification? Must we suppose that language arose by "separate creation"—that it had no roots in earlier preadaptations? Did a fully formed language module just spring into existence by sudden unprecedented mutation? Did human DNA experience a random convulsion resulting in creatures that could speak? Did syntax and semantics emerge from nothing, by a kind of spontaneous generation? Surely we want to avoid that conclusion if we can. But it is hard to see what the preadaptation might be, because language appears *sui generis*: irreducible, a domain unto itself. Our hypothesis is that the hand is the critical variable, but how do syntax and semantics emerge from the hand?

I am not here concerned with the evolution of thought (I will consider that later). I will assume that thought is already in place at the stage of man's evolution we are considering. This was presupposed in the discussion of tool use, since instrumental thinking was taken to be present at that time. Our question, then, is how to get from the mental and physical structures present in manual tool use to language.[1] (I actually hold that thought massively antedates language and is present in many species that do not possess any form of language, but I won't be arguing that position here.) The question is how a communicative system arose from these antecedent conditions, including thought (which I take to involve the possession of concepts). If thought is already present, then *intentionality* is present: thoughts are about things and they ascribe properties to things. My question is how external public events ("utterances") get to be about things and ascribe properties to things; it is not about the origins of intentionality in general. Thus I want to know how the hand could be the basis for reference and predication in a public language, given that reference and predication are already present in thought. It is agreed not to be the basis of reference and predication in thought, since many animals have thought without having hands (e.g., whales). But it might still be the basis of reference and predication in the external communicative system possessed by humans.[2]

We shall not see much hope of an answer along these lines if we insist on viewing language as essentially vocal. How could the vocal emerge from the manual? What has the mouth got to do with the hand? But this is obviously a simplistic and distorted view of the essence of language, because of the existence of gestural language and sign languages such as those employed by the deaf (e.g., American Sign Language). The linguistic cannot be identified with the vocal. Clearly the hand *can* function in a fully linguistic manner, silently.[3] In addition, of course, we constantly use our hands communicatively during vocal exchanges in all sorts of ways. We lose nothing essentially linguistic if we suppose that early human language was (primarily) a gestural language, and there is good reason to think that this was the situation.[4] The question then becomes how a gestural language centered on the hand could arise. What is it about the hand that makes it capable of linguistic structure? What are the preadaptations that enable the hands to refer and predicate? How did gestural language *emerge* from the hands? I am now going to describe three possible theories. They are not mutually exclusive but can be combined to produce a more complex theory; still it is useful to consider them separately to begin with.

(i) *The SVO theory*. Armstrong, Stokoe, and Wilcox believe that syntax can be found in natural actions of the hands.[5] They invite the reader to swing his right hand in front of his body and catch with it his upraised left index finger. This action can be analyzed as follows: the agent of the action is the right hand, the object is the left index finger, and the action is the swinging and catching. They comment: "The grammarian's symbolic notation for this is familiar: SVO" (179). So they see in this simple action of the two hands three components, corresponding to subject, verb, and object. The action "possesses a structure: in it *something does something to something else*, or SVO—the seeds of syntax" (181). Thus "because gestures of this physical type contain the structure of the basic sentence—whether symbolized 'SVO' or 'NP + VP'—they also open the way to a more sophisticated symbol use than naming; they permit language to begin; they symbolize relationships" (181–182). The idea is clear enough: one hand can act as an agent, the other as an object, and an action can be performed by the one on the other. By observing such actions we can abstract a structural relationship that mirrors the SVO structure: we can separate out the components and grasp a relation between them. We have a structured sequence that resembles syntax. If we now interpret each hand as a name of something, the action can be seen as a (relational) predication. So it is not, they contend, that gestures can only attain the level of names; gestures can also function as verblike elements. The act-object relation mirrors the verb-noun relation. We might also cite the way thumb and fingers act on one another to form complex action configurations: the digits can be seen as namelike and the actions as verblike. All the fingers can combine systematically, like words in a sentence, and their actions contain the seeds of predication. The hands themselves have SVO syntax, according to these authors.

This theory is suggestive, if rather underdescribed. The hands certainly have combinatorial complexity, so that we can easily imagine reading genuine syntax into hand actions (as with contemporary sign languages). But the move from action to verb seems precipitous (saltatory)—surely not *all* hand actions are tantamount to verbs. Where does the symbolism come from? Also, the account is purely syntactic; it says nothing of how reference might be grounded in the actions of the hands. Nothing in the prehensive action of the hand finds a place in the SVO theory—the action of hand on object is left out of the picture. We don't get the idea of predicating something *of an object*—just the joining of subject terms, object terms, and verbs into syntactic strings. Indeed, it is hard to avoid the impression that the authors are trading on a kind of use-mention confusion, conflating actions with the verbs that signify them. Can we do better?

(ii) *The grip-action theory.* The heart of this theory can be simply stated: the prototype of reference—its precursor—is the action of gripping an object. Referring emerges from prehending, crudely. Referring can be understood as "virtual prehending." Consider holding an object in one hand and acting on it with the other hand—striking it, scratching it, or rubbing it.[6] The gripping hand functions as the referring term and the gripped object is the reference, while predication corresponds to the acting hand. Here we have the makings of the syntactic structure of reference and predication *and* reference itself as a symbol-object relation. In a subject-predicate structure we have one element that links to an object and another element that applies something to that object: one element "takes hold" of an object, while the other performs the action of predication on that object. Similarly, one hand can take hold of an object, while the other acts on it in some specific manner. The dyadic object-directed structure is present in both.

The grip-action nexus is thus analogous to the subject-predicate nexus. Different actions can be performed on the same held object, as different predications can be made of the same object; and different held objects can be made subject (*nota bene*) to the same action (type), as the same property can be predicated of different objects. Subject and predicate are detachable and recombinable, as object and action are. We take the object in hand in order to do something to it, just as we identify an object in order to ascribe something to it. Predicating is an action, as is referring; gripping is an action, as is acting *on* a gripped object. We intentionally grasp an object in order to act on it, and we intentionally single out an object in order to comment on it. A bit of the world is selected in both cases, so that it should be made the subject of an act. I might pick an object up and act on it to show you how to perform a particular skill, for example, in teaching you how to make a stone chisel. Now there is a social dimension to the action. I chip at the stone so that you can observe how I do it to make a good chisel. Your attention will be directed to the held stone and then you observe the action in order to learn something. Similarly, I can "pick out" an object symbolically, say by pointing, and convey something to you about that object. If my symbol is a manual gesture it will resemble to some degree actually seizing an object in the hand (see below). Complex actions of gripping and acting-on will be common in a social group of tool users, and the grip-action theory sees in this the seeds of language in its most primitive form. It sees a platform (a preadaptation) from which reference and predication might get off the ground.[7]

Note also that gripping an object involves a special relationship between the body and the world, in the form of *isomorphism*. The hand must be shaped *to* the object: the fingers must be so configured that the object is properly held. The shape of the hand is different according to whether a power grip is used or a precision grip—as with gripping an axe versus gripping a writing implement. The fingers must adjust quite precisely to the form of the object (more strictly, its gripped part). Accordingly, the hand must be capable of as many configurations as there are geometrical types of held objects. The varieties of grip correspond to the varieties of objects (and varieties of actions performed with those objects). Thus it is possible to read off the type of object gripped from the shape of the gripping hand, because of the geometrical congruence. In this isomorphism we also see the seeds of representation: the grip is a kind of "picture" of the object gripped, a replica of it, a diagram.[8] It would be possible to use a particular hand configuration as a symbol of the kind of object held by that grip in a gestural sign language. You might form the hand into the power grip used to hold an axe in order to symbolize an axe. Your audience would be familiar with that grip and be able to infer what you are referring to—they might then bring you an axe. The grip is a kind of mirror of an axe, and you can exploit this fact to obtain an axe.

(iii) *The mimicry theory.* In the grip-action theory we see the structure of language in embryonic form, and the way language takes us to the world beyond the body, but we don't yet see anything that deserves to be called actual representation. So the suggested preadaptation needs supplementation, if we are to avoid inexplicable saltation. How does the hand become a symbol *of* something else? How can actions of the hand be interpreted as symbols by observers? Here the answer is not far to seek: the hand is capable of feats of *mimicry*. The hand can *copy* events and states of affairs in the external (to the body) world, and this copying can afford acts of communication. Let us take a very simple example: eating. Eating involves taking hold of food with the hand and inserting it into the mouth (we are supposing early man is a manual eater—no knives and forks). Suppose someone performs this action but without any food in the hand: she is mimicking the act of eating. This act of mimicry might be intended to indicate hunger on the part of the agent, or it might function as an incitement to eat by parent to child. Suppose the parent places food in front of the child and mimics eating it: the child will likely get the message that she should eat the food. Certainly the act of mimicry will bring to mind the action of eating, and then context will supply the intention of the "speaker." Or consider the action described by Armstrong et al.: seizing the left forefinger in the moving right hand. We can easily imagine a context in which this action will be taken to mimic the seizing of a prey animal by a predator—where the prey might be a human. Thus when a member of the group strays toward a certain dangerous area or decides to wander around at night, another member might perform this act of mimicry to warn of lurking predators. The wanderer will (with luck) get the message, because of the similarity between the hand actions and the event simulated. The notable thing about the hands is that, owing to their remarkable versatility, control, and agility, they can mimic extremely well—the feet would be nowhere near as adept. Thus manual icons might develop based on mimicry—that is, standardized gestures that represent types of things.

One form the mimicry might take is assuming a characteristic grip in order to mimic holding a certain kind of object—as with assuming an axe grip in the example given above. I mimic gripping an axe in front of you so that you will bring me an axe. I have thereby referred to an axe. You can infer what I want from the grasping act I have mimicked. In a social group these kinds of imitative actions will have utility, so the ability to mimic will be selected for. The hands clearly have the potential to act as simulations, so we have located the preadaptation we sought. This potential just needs to be exploited in the service of a pressing need. To the enlarged brain, improved hands, use of tools, and social grouping, we can then add mimicry as an accomplishment of post-arboreal man. Man began to simulate nature with his hands, thereby communicating messages to observers. Perhaps he already had some talent as a mimic while still in the trees, as some contemporary primates do, but now he expands this ability, combining it with tool use and more elaborate social structures. The hands will have become more adept and flexible, so the range of mimicry will have amplified. In this we can discern the roots of representation. The stage will then be set for more abstract forms of simulation, and later for hand signs that are purely conventional. From those (not-so-humble) roots the mighty tree of language will eventually grow.[9]

As I said earlier, these three theories are not mutually exclusive; we can combine them. Suppose a member of our early tribe holds in his left hand some found object, say a piece of fruit or a dead bird; then he acts on that object with his right hand in a violent manner, intending to simulate the action of a predator. Here we have the SVO structure described by Armstrong et al., but we also have an analogue of reference in the prehension relation between hand and object, as claimed by the grip-action theory, and we have the element of simulation proposed by the mimicry theory. Thus the action represents a predator attacking its prey by referring to prey and predator and ascribing the action of attacking. Or we might imagine a member of the tribe biting into the fruit and then throwing it in a certain direction, so as to indicate that there is more of the edible fruit in the direction of the throw. Here the plentiful fruit yonder is referred to by holding a sample of it, and the direction of the fruit is indicated by the throw. Reference and predication thus emerge from a combination of structured hand action, manual prehension, and mimicry—syntactic form, objective reference, and iconic representation, respectively. Primitive gestural language thus proceeds on the basis of *three* preadaptations that are brought together organically.[10] But all center on the hands, exploiting their capacity for combinatorial sequential structure, object prehension, and symbolic mimicry. This is quite a rich brew.

In fact, there are two other ingredients to add to the pot. We must not forget that the origin of language proceeds against a cognitive background. First, there is the simple fact that our early humans are already thinkers: they already instantiate the subject-predicate structure in *thought*. So this structure is already installed in the brain; what we don't yet have, prior to the advent of public language, is its manifestation in an outer medium.[11] What we are trying to show is that the hands are suitable for instantiating this preexisting structure externally. The hands thus externalize what is already internally realized; they do not create the subject-predicate structure *ab initio*. So a preadaptation for *linguistic* reference and predication is *cognitive* reference and predication (I would estimate that this goes back at least to reptiles and is premammalian). Second, we have already credited our early humans with specifically teleological thinking in relation to tools—they conceive the world instrumentally. Thus the way is open for them to conceive the hands as tools: tools for manipulating objects, but also as tools for other jobs. They might then see in the hands the potential for a vehicle of communication, given that they are already social beings with a need for communication. They begin to view their hands as *symbolic tools*—devices for communicating. They could then decide to exploit the mimicry potential of hands in communication, seeing this as a means to an end. A stone can be used as a tool for chopping; a hand can be used as a tool for communicating. The tool of language is accordingly constructed, following the general enthusiasm for tools that has seized the human animal since he departed his arboreal home. There is real creativity in this, no doubt, and all creativity is puzzling, but man has by this time long since been an agent of creativity in the construction of even quite primitive tools. In language he has created a new tool, based on an ensemble of resources he already possesses. He puts together these antecedent resources to produce a shiny new implement in the struggle for survival: a system of symbolic communication. Internal thought, bimanual anatomy, digital dexterity, a talent for mimicry, a prehensive lifestyle, a preoccupation

with tools, social coordination—all these feed into the evolutionary process that produces primitive language. This is not something from nothing, but something from quite a lot.

The biological forms that precede and produce language are
therefore multiple, and fall into two categories. On the physical
or bodily side we have the anatomy and functionality of the
hands in relation to the world beyond the body: this gives them
the physical potential to function as a language (unlike, say, the
elbows or the feet). They are free to be used at will and are capable of fine discriminations of movement. On the mental side we
have a rich array of psychological capacities that can feed into
language production: we have structured thought, intention,
teleological rationality, tool consciousness, social coordination,
and creativity. The suggestion is that if we join these two categories together we can begin to see how language might intelligibly emerge—not from a void by miraculous saltation, but
from a rich and complex set of preadaptations. The transition
is therefore smooth, not abrupt; incremental, not revolutionary
(of course, the *effects* of language can be revolutionary, even if
the origins are not). Man needs merely to apply his burgeon-
ing tool-oriented intelligence to the physical capacities of his
hands—they are the ideal means to the end he seeks, and they
are right in front of his nose. He needs to communicate with
others in his new cooperative lifestyle, and his hands are the perfect organs to be so employed. They *already* have the structure
and functionality needed to operate as a medium of linguistic
communication. Thus, in sum, the hands graduated by incremental steps from brachiating to tool using to talking.[12] Talking
was *latent* in the hands—a talent just waiting to show itself. The
hands had the capability to act as sentences; it was up to their
owners to exploit this capability—and they had the intelligence
and desire to do just that.

7 Ostension and Prehension

In the previous chapter I made a comparison between gripping and referring. It may seem that this is more metaphorical than literal; after all, you do not (usually) physically grip the thing you are referring to. In this chapter I shall respond to this worry by presenting a theory of ostension based on prehension. I shall argue that ostensive pointing can be interpreted as a kind of extended gripping: it is like gripping with an extended virtual hand. It is a kind of imaginary gripping. Gripping is the preadaptation that imagination augments into ostension.

Consider the pointing gesture: the index finger protrudes stiffly toward the object of reference, while the thumb (including its fleshy base) closes over the remaining fingers, which are tucked firmly into the palm. The result is a partially open grip, with one finger extended and the rest closed as if holding something. We do not point with a fist using the index finger's knuckle as pointer; nor do we open up the fingers and orient them all in the direction of the ostended object. The fingers are partly open and partly closed. What are we to make of this?

We can grip something in the hand so as to conceal it, in which case the fingers close over the object, hiding it. Or we can open up the grip to reveal what is in the hand, partially or totally; a semiclosed grip is advisable if the object might drop or spring free.[1] We can call this the *display grip*: it is useful in showing to others what is in your hand. The act of showing is an unfurling of the fingers, wholly or partially. We can envisage early humans using a display grip when showing food held in the hand to infants—nuts and berries, perhaps. Closing the fingers and palm over the object would be typically used in the *carriage grip*—the grip used when transporting objects from place to place. There is also the grip we use when giving an object to someone else, where the identity of the object is revealed, but the object is still partially gripped to avoid slippage. Call this the *transfer grip*. Imagine an ancient human using the transfer grip to give food to an infant or partner, with the recipient observing the form of the grip. This will be typically a semiopen grip, somewhat like the pointing gesture, in fact. But suppose the parent does not actually have the food in her hand, yet wants the infant to take the food. She might physically touch the food she wants the infant to take, in a kind of lazy gripping action, instead of picking it up. Instead of grasping the fruit and handing it to the infant, she briefly encloses it between thumb and forefinger, indicating that she wants the infant to eat the fruit (she is tired of putting it in the child's mouth herself and wants him to learn how to feed himself). In so doing she directs the infant's attention to a particular object in the environment. Or she might just push it toward the infant using her fingers. She uses what we might call a *diminished grip*. Thus the robust transfer grip naturally evolves into a more gestural diminished grip.

But we can imagine a further diminution or attenuation: the fruit is too far away to touch with her fingers, so the mother picks up a stick and uses it to touch the fruit in question. Maybe in the past she has used a stick to impale the food and physically insert it into the child's mouth, so this is not much of a departure. Now we have a *prosthetic diminished grip*. Still the child gets the message: "Eat that piece of fruit!" But what if there is no stick around? The mother might then extend the index finger in the direction of the fruit, so that it is *as if* she is touching it. Finger and fruit might be only inches apart, or the distance might be greater, depending on the context. After a while the gesture might become stylized, with traces of the original holding grip still present, but the index finger extended out as if to touch the object. The gripping has become virtual—merely imagined. Thus the transfer-display grip has turned naturally and gradually into what we think of as the pointing gesture. It has become a notional gripping, a virtual prehension, an imaginary touching.[2]

Something of the evolutionary history of the ostensive gesture, so conceived, apparently survives in current practices. On occasion we are not content with distant pointing, perhaps because of possible ambiguity; we might then walk right up to the object and jab our forefinger into it. Thus the army sergeant approaches the delinquent cadet and jabs his forefinger into the lad's chest, exclaiming, "You, yes *you*, report to the captain at once!" Or one might walk down a line of people all wishing to be chosen for a team and lightly touch a subset of them with the tip of one's finger, thereby indicating who is to join the team. We thus reverse gestural evolution, returning ostension to its grip-and-touch origins. It would also be possible to briefly grip each candidate's shoulder, as a gesture of inclusion in the team—this would serve much the same purpose as pointing the person out from a distance. Also, we often resort to the use of a "pointer" when engaging in certain kinds of ostensive reference—a specially made rod used to touch, say, particular parts of a map about which one is discoursing. Distant pointing with the finger may be too crude and ambiguous, so we extend the pointing finger with a suitable prosthesis. Again, the act of reference has strong connotations of touching and gripping (the pointer itself is gripped).

There is thus a continuous sequence from gripping in the hand to displaying with the hand, to transferring, to touching, to using a pointer, to pointing from a distance. Preexisting forms of hand action are gradually modified to serve a felt need. Ostension has its preadaptations and precursors in antecedent manual acts.[3] The shape of the ostensive gesture, as a partially open grip, is testament to its history—we have here an instance of ancestral preservation. As a matter of logical possibility, a species might point with the elbow or the tongue or even the rump; but the way humans point reflects the roots of the gesture in prehensive acts of the hand. Human pointing is prehension modified.

We can also note that the ostensive gesture characteristically involves extending the arm; we do not point "from the hip." Thus the action simulates reaching for something, which is a prelude to seizing and grasping it. So the whole action looks like attenuated reaching—truncated reaching, we might say. Ostensive reference to an object resembles reaching out to the object. If we had indefinitely extensible arms, capable of reaching and seizing any object in the environment, we could dispense with remote pointing, and simply extend our hands to touch the object of reference.[4] We can certainly imagine a species with special tentacles that never points without actually touching the object of reference, even gripping it with suction pads for the duration of the act of reference. Our fingers are comparatively short and inflexible, so we replace touching with virtual touching, gripping with virtual gripping. The imaginary line projected from the tip of the pointing finger to the object acts like an extension of the finger, but we can conceive of fingers that never need such virtual supplementation. What we call ostensive reference is touching gone virtual, according to this way of looking at things.[5]

It is instructive to compare the pointing gesture with the so-called come-hither gesture. When beckoning someone, it is common to pull the extended fingers back into a semigrip repeatedly. Compare this with actually seizing the person and gripping him. The come-hither gesture mimics gripping another person and pulling him in. We can see how it might have developed from actually seizing someone, becoming attenuated and stylized. It comes to symbolize what it used to do physically, and is interpreted that way. A person not wishing to accede to the beckoning gesture might extend an upraised palm in a kind of pushing motion, as if resisting the other's attempt at gripping. We have virtual seizing and virtual resistance. Another example might be the "farewell gesture"—waving goodbye. This action looks a lot like stroking or grooming, and we can easily imagine that it arose by attenuation of actual stroking. It becomes a ritualized gesture of affection or solidarity.

In the light of these examples, we might postulate a *law of gestural attenuation*: the tendency of a hand action to become modified into a weakened or diluted version of its original, thereby becoming more symbolic than actual. We thus get "action at a distance," as the original action is performed at some distance from the object of the action, in stylized form. If this is on the right lines, then the ostensive gesture is one instance of the law of gestural attenuation—it results from a weakening of an original action of gripping and touching. In all these cases the derived gesture morphologically resembles an action of the hand that involves actual contact. It is a bit like the "air kiss": it resembles a real kiss, but is performed at a safe distance. The "fist pump" is similar: at close quarters it would be a punch in the stomach, but from a distance it signifies victory or superiority or domination. The actions become physically detached from their object, but they mimic the original action quite closely.[6]

Thus reaching and gripping become modified into virtual prehension, and this is what the ostensive gesture amounts to. Since the ostensive gesture is a (or the) basic mode of reference, we can begin to see how reference evolves from more primitive prehension (combined with the other factors I mentioned, such as imagination). Pointing to something is a modification of a preexisting form: the action of gripping (and therefore touching) an object. If we think of the original grip as a precision grip, the kind used in holding a nut, say, then the index finger is already assuming the posture of pointing in embryonic form: the nut will be delicately gripped between thumb and partially extended forefinger, with the tip of the forefinger pressing lightly on it. Releasing the grip, as in handing the nut to someone, involves extending the forefinger slightly, so that it becomes straighter. This is the posture of the finger that is used in merely touching the object, as when the mother indicates to the child what to eat; and then increasing the distance between finger and object has the form of the pointing gesture. The transition is smooth and natural: there is no sudden saltation whereby a chance mutation gave rise one day to an animal with an unprecedented ability to point at things. Pointing has its evolutionary prehistory, its precedents and precursors, its embryonic form—following the general laws of ancestral preservation and incremental adaptation. Just to have a convenient (and catchy) label, we might call this the *haptic theory of reference*, because of its emphasis on the role of *touching* in the genesis of reference—though this oversimplifies the full story.[7]

Let me end this chapter with a methodological disclaimer: what I have suggested here, as elsewhere, is highly speculative. I have no direct empirical evidence that anything like the story I have told actually occurred. It is also only the beginning of a complete theory. The point is that such a story renders the evolution of ostension intelligible—it enables us to see how ostension *could* have evolved. This is what *may* have happened. That is, it is a theory—a hypothesis. I am not asserting that I know this theory to be true, only that it *might* be true. It is coherent, explains the data, and respects well-motivated constraints on evolutionary explanation. It is a candidate for truth. And I don't know of any better theory. That is all I claim—not some sort of magical insight into what happened in the remote past, or hard fossil evidence demonstrating that pointing evolved from prehension. (I make this disclaimer in response to stern "scientific" critics who accuse me of claiming more than I am evidentially entitled to, as if I am indulging in some sort of methodological delinquency. That is to miss the point of what I am doing.)

8 From Signs to Speech

A manual sign language is silent; speech is noisy. The larynx produces the latter (along with other vocal organs); the hands produce the former. How did one evolve into the other? For what reasons did the transition occur? If we agree that the original language of humans was (mainly) a manual sign language, then what prompted us to switch to a vocally based form of language? Is a sign language limited or unsatisfactory in some way—a mere "starter language"? Sign languages as they exist today, as in the languages used by the deaf, are highly iconic and indexical, whereas spoken language is hardly iconic at all and less dependent on context.[1] Does this make sign language defective by comparison? Not obviously. Another difference is that sign languages are parallel whereas spoken languages are sequential: the hands can perform several gestures (words) simultaneously, but the vocal organs can only get out one sound at a time. This seems like a *dis*advantage in vocal language. It also seems more effortful to speak using sounds, and the vocal organs can quickly become fatigued; by contrast, the hands can continue signing indefatigably, more or less. Moreover, visible signs can operate over longer distances than spoken sounds: you can see gestures from further away than you can hear vocal sounds. Then there is the point that ambient noise will disrupt vocal communication, but not visual communication. So why are we not still speaking in signs? Wouldn't this be preferable? The fact that we are not makes one question whether the gesture theory is on the right track after all. Clearly, we need to find out why vocal language became more adaptive than sign language as time passed.

As language became more entrenched in the early human population, which probably took thousands of years, the more time in their lives it took up. Humans went from laconic to loquacious (so it may reasonably be supposed). As they became more complex socially, more interactive, the need to communicate grew. Organizing hunting trips will have required more communication, and the cooperative making of tools would also call for increased conversation. Assuming they used a sign language, that means the hands were increasingly occupied with matters linguistic. But that posed a problem: if the hands were being used to communicate, they could not perform other actions at the same time.[2] They were, to be sure, not as preoccupied as the feet, but they had their jobs to do too—chiefly, the making and use of tools. You can't easily communicate with your hand if you are holding a spear or an axe in it, or chipping one stone with another. Carrying anything and talking at the same time becomes impossible. Thus we have the problem of multitasking. The hands became so important in the expanding tool world, and perhaps in other respects too, that it was impractical to use them for communication as well. You don't want to have to put your precious tool down just so you can make a remark (think of issuing orders in the heat of battle). It therefore became advantageous to communicate through some other channel. We can assume that the vocal organs were already being employed for some measure of communication, as they now are in apes, so they were adapted for communication already, if of a rather limited and crude kind. So selection pressure began to apply to the vocal organs, forcing them to acquire greater range and skill over the generations. Most likely a mixed system was the norm for a while—signs and sounds having roughly equal roles. But the more the task of communicating was carried out by the voice the more free time the hands had for other vital work. Hand time was too precious to waste. Thus sign language became gradually phased out—though not totally, because we still use it today, in vestigial form, but at least as the primary vehicle of communication. A hybrid system of gesture and sound, with sounds substituting for gestures over time, eventually became the norm. This freed the hand from communication duties, as it had once been freed from tree-climbing duties. Simultaneous talking and manipulating was the big payoff, still very much in evidence today. So, at least, it seems reasonable to suppose (see the methodological disclaimer at the end of the last chapter).

Subsidiary problems arose with the use of visible signs. People have to be looking at you if you are to get your message across, and they cannot look at many speakers at the same time. Obviously you cannot communicate with someone this way if you are behind a bush or otherwise hidden or even turned away. What if someone steps in front of you mid-sentence? This problem of visibility is acute at night, when darkness will obscure manual gestures. If humans became more nocturnal, perhaps hunting at night, this would have been a real drawback. You cannot be seen, so you cannot be heard. Clearly, a vocal system will work much better in conditions of darkness. It would be possible in principle to use the hands to produce sounds, but this is very limited as a system of communication—clapping will only yield so much semantic content. Drumming can be used in certain settings and is still used today by some percussion-oriented tribes, but carrying a drum around with you all the time is a nuisance. In addition, a problem of privacy arises in sign language: it is hard to *whisper* with your hands. As social relations became more complex, and perhaps more calculated, it would be desirable to be able to communicate without being overheard; but gestures can be seen by anyone in the vicinity. With the voice, however, it is possible to moderate the volume and move in closer to the hearer's ear, thus preventing eavesdropping. When it comes to plotting and gossiping—two notable human traits—the voice has a distinct advantage over the overt gesture.[3] These factors would also impose some selective pressure in the direction of a vocal language. What we would expect, then, is a multimodal communication repertoire, with hands and eyes, and voice and ears, all involved—and this is exactly what we find in modern humans. This will have evolved out of a predominantly gestural language, by incremental stages. Indeed, it may be true that a vocal language *could* not have evolved in any other way, given the adaptive realities of humans; it *had* to proceed from a largely gestural initial stage. Manual language would then be the necessary preadaptation for vocal language.

It is easy to exaggerate the differences between visible signing and audible speaking because of the sensory difference. But examined from a motor point of view the difference is less pronounced. The vocal organs produce sounds precisely by forming particular shapes of the larynx, tongue, lips, and so on. Air is expelled from the lungs through these shapes, and sounds are thereby produced. But the underlying cerebral apparatus is essentially an engine for producing configurations of the body— just as with sign language. The tongue is functioning in much the way the fingers do, moving and forming discrete shapes as we speak. The tongue is a fingerlike prehensile organ in its way, so the brain controls it much as it controls the fingers of the hand. The hand can make sounds by moving in certain ways, and the tongue does the same, aided by the lungs. It is quite possible to imagine an apparatus, either evolved or invented, that attaches to the hands in such a way that sounds are produced when the hand assumes particular configurations. Then the hands could be part of an auditory system of communication. Likewise, we can imagine a species that talks with the tongue but makes no sound—instead, observers *look* at the tongue of the speaker and infer the speaker's intentions. They substitute tongue shapes for tongue sounds. The difference between signing and speaking is thus rather adventitious, conceptually speaking. It is by no means impossible that the brain structures underlying manual language were co-opted to power vocal language—since the latter is just as "configurational" as the former.[4] Tongues are just contingently more hidden than hands, but they assume complex shapes as much as hands do. The brain machinery controlling speech may therefore have had a preadaptation in the machinery already existing to control gestural language. Certainly abstract structures, such as subject-predicate forms or quantificational forms, can be common to both sign language and vocal language. The abstract structures of language transcend their specific sensory manifestations. The transition from gesture to speech is therefore not as dramatic as it might appear, once we focus on the motor side of things. Speaking is making movements, whether these are seen or heard, whether by hand or mouth.

If human language was originally a gestural language, centered on the hands, writing is not as "unnatural" as some may assume; in fact, it is closer to primeval language than contemporary speech. What is sign language but "air writing"? The signing hand traces word shapes in the air, as the writing hand traces word shapes on paper. If the Earth's atmosphere had been different—thicker, more viscous—there might have been traces of gestures persisting after the gesture had ceased, and then signing would *look* more like writing. When we use the hand to write we are quite possibly tapping into ancient brain circuits that served primitive sign language (and still serve the gestural elements of language that we use today). We might, somewhat romantically, view writing as closer to our original hand-centered symbolic nature, not as a newfangled technology that goes against our basic vocal instincts. Speaking is the recent innovation; writing has been around ever since language began—even if only as "air writing." The hands and language have been intertwined for hundreds of thousands of years (or for however long language has existed—which we don't know).[5] When you sit typing at your keyboard or pen a letter in the old-fashioned way, you may be closer to ancient man than you realize, as he sat around his campfire gesticulating to his fellows with wild enthusiasm. Primeval man was a *writer*.

If these speculations are along the right lines, then human language has a fundamentally different origin from nonhuman languages, such as those of whales and dolphins. These species have no hands and do not use sign language (or not much—their fin movements might have some communicative significance): their language is wholly auditory. It would be highly implausible to suggest that whales and dolphins went through an early evolutionary phase in which their language was gesture based and only subsequently became sound based. Whales and dolphins *have* no ancestors with hands like ours. Their languages must have been auditory from the start (though the idea of a simple language based on smells has been mooted). But if so, the evolutionary trajectory of these nonhuman languages is not like that of our language, which passed through a (largely) non-auditory phase, according to the gestural theory. What we have here is a case of evolutionary convergence: humans, whales, and dolphins have converged on auditory language from different starting points, and quite independently. This is due ultimately to our arboreal past (whales presumably had no arboreal ancestors, despite being mammals). The three species do not have a common ancestor with an auditory language from which we all descended. Evolution "discovered" auditory language twice or more, presumably because of its distinct advantages (for whales and dolphins sounds travel better than gestures in the murky depths of the sea). The *reasons* humans and whales came to use an auditory language are quite different, stemming from these species' very different environmental niches and needs. It is not that naturally evolved languages are *essentially* aural, or even that this is their ideal mode of existence. Language as such is modality neutral.[6]

According to this historical picture, then, humans came by intelligible steps to be both a vocal species and a gestural species. First we were gestural; then we became vocal (to oversimplify slightly). We are conspicuously a species of talking toolmakers in our modern form. Our characteristic mode of being and behaving is simultaneously to talk and to use tools. Our hands are always busy with our "equipment" (as Heidegger designates it[7]) and we concurrently comment on sundry topics. We eat with cutlery and chat; we drive cars and converse; we play musical instruments and sing. Our hands and tongues are always happily gyrating together, always simultaneously active. At one time we (or our close ancestors) were relatively silent creatures and our hands were just one bodily organ among many—this was during our untroubled arboreal phase. But once ejected from that habitat we came by a series of steps to be hand-centric and ceaselessly verbal. As terrestrial animals, we became the upright large-brained tool-obsessed big talkers we are today. Extraterrestrial zoologists visiting Earth would no doubt be immediately struck by our nimble digits and our voluble stream of speech. The precondition for this was our expulsion from the trees and the necessity for compensatory adaptations.[8]

9 Hand and Mind

I have suggested that the hand played an essential role in the origin of language (as have others), but I have not claimed that the hand generates the mind. Those who believe that mind and language are coeval, or even that language is the basis of mind, might make that very strong claim: that the hands are the preexisting form from which mind evolved. I do not find such a view remotely plausible, since I think many animals without hands have minds—such as dogs, cats, and elephants. These animals have paws and hooves, but they don't have hands. I also accept that reptiles, fish, and even insects have minds, even if of a rudimentary form: so I am not about to suggest that hands are a necessary condition of mindedness. But there is room for a weaker thesis about the human mind in relation to the hand—namely, that the *form* of our mind is shaped by our handedness. The *kind* of mind we exemplify is influenced by our possession of hands—hence our specific type of mind evolved from conditions that include our hands. In this chapter I will take a brief synoptic tour of this terrain, without attempting to treat the several topics in any depth. A great deal could be said about each of these topics (and a lot has been said). My aim is just to indicate the many ways that the hand can be seen as shaping the mind, so that we gain a sense of the pervasiveness of the influence. Accordingly, I shall be brisk.

What is most distinctive about human intelligence, as has often been remarked, is that it is a tool-using intelligence. Thus it is creative, teleological, and manipulative. We are inventive creatures; we conceive the world instrumentally; and we confront reality with a view to changing it. We don't just accept nature as it is; we actively interfere with it. We try to improve on nature, to serve our own ends (often in ways deadly to other species). Large tracts of the material world are thus transformed into human equipment. Other species accept nature as given, with some very minor tinkering, but we engage in wholesale reconstruction. (It is true that some other species are also tool-users, but we have taken this trait to an extreme that far outstrips even the most tool-oriented ape.) We are surrounded by a world of our own making—a world of technology.

The intelligence that makes this world is grounded in the hands, because almost all technology is hand operated: it is designed to fit the anatomy of the hand and the hand's manipulations control it. We could not have created this world of tools and technology without the hand and its peculiar powers. Our specifically human intelligence has evolved in concert with the hand—so much so that we can speak of "the intelligent hand." But we also, and equally, have a "handy intelligence." We express our intelligence in our hands (also in the voice—but language came from the hands originally). What we are particularly intelligent *at* is figuring out how to make and use hand-operated tools that solve problems.[1]

We may conjecture that the hand plays a major role in the growth of individual intelligence, as well as the growth of species intelligence. The child uses her hands to manipulate and explore, acquiring hand skills of many kinds, and without this manual activity intelligence would no doubt be impeded. It is an empirical question to what degree intelligence is affected by hand activity in early years. Those born without hands, or with impaired manual dexterity, might be expected to suffer some intelligence deficit (the same would not be true of the feet). Of course, merely lacking hands does not entail that the parts of the brain dedicated to the hands are also missing or defective, so general intelligence might still be grounded in this. But I would guess that if anyone were born without this part of the brain then intelligence would be severely affected. (I do not know of any studies that provide good data on this question, possibly because that kind of massive but selective brain impairment is very rare and likely to be conflated with other deficits.) A lot of cortex is devoted to the hand, so damage to that part of the brain is bound to have a considerable impact. Lacking even neural programs for manual activity must surely affect intellectual maturation (by contrast, being born without a sense of smell is unlikely to have much of a deleterious effect on general intelligence). When psychologists speak of sensorimotor intelligence, it is primarily the hand they have in mind.[2]

The hand is also one of the main ways we *learn* about the world—it is a sensory organ as well as a motor organ. It is hard to believe that this fundamental mode of learning does not shape the way we conceive of things. Our concepts of things will naturally incorporate the way those things are presented to the probing hand. We don't just have visual and auditory ideas of things; we have manual ideas. A cup, say, is an object that is to be gripped thus-and-so. If we think of perception as incorporating a motor component (as with *enactive theories*[3]), then the actions tend to center on the hand; and manual perception feeds into our concepts of the world. Our conceptual scheme is accordingly geared to the human grip (though not only to this). The real is what can be gripped, to put it succinctly; the unreal is what cannot be gripped. Thus our ontological preference is for middle-size dry goods—those we can get our hands on and around. Numbers, minds, and values elude our grip, so we fret about their ontological status. We are instinctive "manual realists": we believe most firmly in what can be held in our hands.[4] Other species might be "smell realists" or even "echolocation realists": the world is what can be smelled or detected by rebounding sounds. We humans interact with the world mainly through our hands, so we tend to favor their take on reality. If we are doubtful about a visual impression, suspecting illusion, we reach out to see if we can grasp the suspected figment. If you have something firmly gripped in your hand, you don't doubt its existence. The hands are hard to fool—there are not many manual illusions (compared to visual illusions).[5] Thus human conception and human prehension go, as it were, hand in hand. What would it be like to hold a round square? Ghosts cannot be grasped. Those elusive quanta seem hard to get a grip on. Epigrammatically: the human hand is the measure of all things, so far as human ontology is concerned.[6]

Then there are the archetypal ten digits. It doesn't take much ingenuity to see a link between arithmetic and human fingers. We count on our fingers; we use base 10; fingers and numbers are both called "digits." Our elementary mathematical thinking is surely influenced by the structure of our hands. Perhaps in the dim, distant intellectual past it was the hands that gave rise to mathematical ideas. And not just arithmetic, but also geometry: the closed figures—triangles, rectangles, and circles—that can be formed with the fingers, as well the lines on the hands, and the ability of the index finger to trace shapes in external substances like mud or sand. This influence may go beyond arithmetic and geometry and extend to our digital mode of intelligence: we tend to conceive things in tidy discrete units that combine to form wholes—particles, words, and quantities of all kinds. Is this a reflection of our digital hand anatomy? Did we learn this mode of thinking from our hands? A species centered on smell may approach the world in a more analog fashion, because smells are not discrete combinable units. But we apprehend things via the discrete articulated structure of our hands—thus we tend to think digitally. Ten seems to us like an especially distinguished number, the most "natural" of natural numbers. And we like to think in terms of finite combinations of neatly bounded entities that work harmoniously together. The digital hands are our model and paradigm. If you are a snake or a jellyfish, your take on things will be different: perhaps the world will seem more amorphous or seamless—more like you. The mind of a dog, say, will likely reflect its jaws—the *teeth* will be the dog's conceptual template or point of ontological reference. The world will be the bite-able world, with a basically binary structure (corresponding to the upper and lower jaw: here I venture into the canine *Weltanschauung*).[7]

Phenomenology and intentionality will thus reflect the anatomical and physiological properties of a species—how could they not? So there will be variation in the general form of an animal's mind, according to its bodily makeup: phenomenology reflects phenotype. In humans the hands are vital to our interactions with things, so their form is likely to shape how we mentally represent things. One's own body is the paradigm physical object, the primary reality, so everything else is apt to be conceived in the light of its peculiarities. An alien made entirely of gases or pure light would have its own distinctive ontological take on the world, quite different from ours. A supernatural being composed entirely of numbers (here we contemplate the counterpossible!) would likely have a different ontological slant again, in which perhaps middle-size physical objects appear as distinctly shady.[8]

I have just explored, all too briefly (and jauntily), some ways in which the mind of man might be shaped by his anatomy, specifically his hand; but there is also the question of whether our *concepts* of mind might be shaped by our concepts of the hand. Do we conceive of the mind in manual terms? Here the evidence is clear and immediate. Our language is full of prehensive terminology for the mind: "apprehend," "comprehend," "grasp," "be gripped by," "pick up on," "hold" (in memory), "catch" (your meaning), "grapple" (with a problem), "reach out" (emotionally), "be seized by" (a passion). Let us take the verb "grasp" as representative of this list; it is common and entrenched. We can be said to grasp a meaning (a sense), to grasp a theory, to grasp an implication—just as we can be said to grasp a doorknob or a tennis racquet. In philosophical jargon, we can be said to grasp "intensional entities" as well as physical objects. Nor is there any sense of outright ambiguity here, as if we just happen to use the same phonetic form for unrelated things. No, the idea is that the two "grasping" relations are significantly similar. We think of the mind as doing to a meaning what the hand does to an object—seizing and grasping it. We model mental grasping on physical grasping—as if a mental hand reaches out (later I will discuss how literally we can take these ways of talking). There is indeed an established philosophical use of "prehend" to describe mental acts like perceiving and thinking. C. D. Broad uses "prehend" in place of Russell's "acquaintance": where Russell says that the mind is acquainted with a particular, Broad says that the mind prehends a particular.[9] The word comes from the Latin *prehendere* which simply means "to grasp." The *OED* gives the following as a definition of "prehension" in philosophical discourse: "an interaction of a subject with an event or entity which involves perception but not necessarily cognition." This fails to capture the active nature of prehension: the mind "takes hold" of the object, grasping it—it acts *on* the object. This is a quite different conception from theories that model intentionality on causation: here the object acts on the mind and the mind is passive. But when an object is mentally grasped, it is the mind that is agential. Or perhaps we should say that the *person* grasps the object (via the mind), to avoid a potential category mistake. The person is the agent of a prehensive mental act, directed toward an object. The naturalness of this way of talking reflects the influence of the hand in shaping how we think about the mind. We grasp the mind *as* a grasping organ—we apprehend it as apprehensive. Clearly, our psychological concepts are saturated with prehensive imagery.[10]

A final point about mind and hand concerns *expression*. Certain parts of the body are held to express the mind with particular force or directness (not counting speech). Thus the face is taken to express states of mind with special clarity—we speak readily of sad or joyful expressions (frowning, smiling), expressions of surprise, of agreement, of sympathy, and so on. A person's face can express inner tension or relaxation, according to the disposition of the facial muscles. Not so the back or thorax. But the hands too are highly expressive: hand gestures can be sad or happy, express agreement or sympathy, and show tension or relaxation. Gestures can convey affection or dislike of another. The handshake is particularly expressive. The mind finds a natural outlet in the hands, as a mode of externalization. Our picture of an emotion may be hand based—as with a gesture of resignation or defeat. The raised hand of victory is understood by all. In the pictorial and plastic arts representations of the hands are often highly expressive. We may wonder whether our emotions would be quite the same without hands to express them (as the smile seems so integral to joy or amusement). Thus the emotions of different species may be shaped by the expressive capacities of their bodies: think of the wag of a dog's tail or a cat's flattened ears. Obviously, too, the hands are implicated in musical expression and dance, and in expressive writing and painting. The hands are the emissaries of the soul, we might say, as the eyes are the windows. You certainly don't need hands to have emotions or a soul, but the kind of emotions you have cannot be detached from the kind of body you have. The emotional repertoire of humans, especially the social emotions, has been forged in tandem with the evolution of the characteristics of the human hand. The hands express human emotions, but they also give them their shape.

In conclusion: the hand is not the very origin of mind, to be sure, but it does have important constitutive links to the kind of mind we have evolved. The specific form of the human mind is to a large degree structured by the human hand, by its anatomy and uses. This is not surprising, given that our minds and hands have coevolved over many thousands or millions of years of biological adaptation.[11]

10   Selective Cognition and the Mouth

In the previous chapter I accepted that possessing a hand is not a necessary condition of possessing thought. The reason is that many living animals have thoughts but no hands. It is not a biological law that thinking animals are handed animals. Thus thought must be an evolutionarily accessible trait for an animal independently of any preadaptation in the form of a hand. But this admission does not rule out a theory of the origin of thought in which *prehension* is crucial, because the hand is not the only prehensive organ. The mouth is also prehensive, and it is far more widespread in the animal kingdom than the hand. I want, then, to consider the idea that *oral* prehension might lie at the root of thought: that is, the mouth was the preadaptation that gave rise by intelligible modification to thought. The mouth is what made thought possible, by providing a platform from which thought could evolve.

Let me emphasize at the beginning that this is a highly speculative proposal, which I do not expect to meet with instant acceptance. I am myself by no means confident of it. But I think it is a useful theory to consider, because it illustrates the kind of evolutionary question we should be asking: namely, how can a certain trait have intelligibly arisen? The constraints on explanation are quite clear and sharp (we must eschew salta-tions): we need to find a preadaptation for every adaptation. So what preceded thought in the biological world that could serve as a preadaptation for thought? Thought had to come from something—but what exactly? At least the theory I shall con-sider makes an attempt to answer that question.

I must immediately make two terminological disclaimers. The first is that I do not intend to explain *all* features of what we call "thought" this way: I do not intend to explain the con-sciousness of thought via the mouth, nor the logical form of thought, nor its role in reasoning, nor its connection to other psychological phenomena. I intend to explain only *one* feature of thought, which I call *selective cognition* (it could also be called *cognitive selection*). By this I mean the ability of thought to con-cern a single object in abstraction from surrounding objects: to pick one object out and focus on that object to the exclusion of others. In sensation, many objects are presented simultaneously, as parts of a manifold; in thought we can select one of these objects, ignoring the rest, and predicate things of it. There is a singularity that thought possesses that is not possessed by sen-sation. Psychologists speak of "selective attention"; I am speak-ing of something similar—the ability of thought to select one object as cognitive focus. Instead of being assailed by a plural-ity of impinging objects in sense perception, the organism can choose one object to think about. I assume that this basic cog-nitive ability evolved well before the full capacity we now call "thought," which brings in many other features; indeed, it may have evolved originally in creatures now extinct (like so many existing adaptations). So think of some ancient creature of the deep that first possessed the cognitive ability to select—I am try-ing to explain how *that* creature acquired the ability in question.

The current human ability to select cognitively will have evolved from this very early adaptation, but it might be as different from the first form of the ability as the human eye is different from the first eye (from which our eye is also descended). Our selective cognition descends from the earliest form of the capacity, but the modifications and refinements will doubtless be enormous. We might refer to the early form as "proto-thought" or "proto-cognition," if we are squeamish about applying the usual terms to these relatively primitive creatures.[1] In any case, I take it that a kind of mental selection evolved long ago in quite primitive animals (though ones capable of sensation)—and I want to explain this without postulating saltations or ungrounded leaps forward. I want to identify the evolutionary precursor. That is, I am asking what needs to be added to basic sensation (which may be regarded as a form of consciousness or sentience) in order to derive selective cognition.

The second disclaimer is that I do not mean any old mouth. I do not mean just an aperture through which food passes into the animal's digestive system. The mouth must be significantly prehensive—some sort of hinged affair capable of seizing and gripping an object. It need not have lips and teeth, but it should open and shut and close around objects. It should be able to take in some given object to the exclusion of others, as with food selection. The idea, then, is that selective cognition evolved from oral prehension—loosely, "thought came from the mouth." I will argue for this theory by presenting the formal features of selective cognition, as I understand it, and then comparing them with the formal features of oral prehension.[2] Then I shall make a suggestion as to how the former developed from the latter.

The formal features of thought (selective cognition) are as follows: object uptake, object retention, cessation of object retention, reuptake of object, object switching, volitional character, internality, and mode of presentation. Thus a person might start to think about an object at a given time, keep thinking about it over a period of time, stop thinking about it, start thinking about it again, switch to thinking about another object, and do all this as a matter of will, with the process proceeding "internally" ("in the mind") and so that the object is presented in a specific way to the person. Bleached of the irrelevant current connotations of the word "thought," we can say that primitive selective cognition involved cognitive uptake, retention, cessation, reuptake, switching, agency, internality, and mode of presentation (perceptual). An object is selected, dwelt upon, dropped, brought back, ignored in favor of another object, and so on. The process is a dynamic and purposeful direction of cognitive resources onto one object at a time, abstracting away from the sensory manifold. The mind "homes in" on a single object, from among many, and holds it in "thought" for a while. It is important that the process is purposive, because thinking about an object is subject to the will but perceiving an object is not.[3] One of the things we are trying to explain is how thought differs from perception in its object-directedness: thought is active and willful but perception is passive—I can decide what to think about but not what to see (given that my eyes are open, etc.). So our creature of the deep is capable of these active feats of cognitive selection—it is not simply at the mercy of what floods in through its indiscriminate senses.

It is easy enough to see that the formal features of oral prehension mirror those of selective cognition. The organism is surrounded by an array of objects, of which it has sensory awareness, but it purposefully selects only one object to hold in its jaws (imagine this is one clam chosen from a bunch of other clams). It holds this object in its mouth for a while, perhaps tasting it; then it may release the object. Later it may grip the object in its mouth again, or it may grip another object. All this is done purposefully. The object is inside the animal's mouth for the duration of the gripping ("internality"). Also, the object will be perceived in a certain way while in the mouth—a sensory mode of presentation will accompany the gripping. The object is temporarily inside the animal's oral cavity, presenting itself in a certain way, as a result of the animal's purposes. Dynamically and structurally, then, oral prehension mirrors cognitive *ap*prehension. Abstractly considered, the processes are notably isomorphic. Thus, if you wanted to construct a new trait with a certain structure on the basis of an already existing trait, this would be a good way to go. The basic design features are already present; you just need to make certain modifications and upgrades. We have a kind of mind-mouth isomorphism, so the one *could* be the launching pad for the other. We can build the new structure around the design of the old structure, because the basic engineering specs are the same. Since evolution tends to be economical and conservative, this conversion of function will appeal to it. We have here a possible case of *exaptation*, the deployment of an old trait in a new setting.

But how exactly does the mouth, an external physical organ, get turned into a "mental organ"? That looks like a big transformation, a saltation no less. It is not as if the mouth became progressively smaller over evolutionary time, until it vanished and transformed itself into the mind! Nor did the animal's brain develop a region with actual prehensive powers—a weird brain-mouth. Where are the missing increments in our story? Where is the continuity we seek? To answer that question, we must turn our attention from the mouth itself to the brain machinery behind it. The brain machinery must somehow code the actions of the mouth—it needs suitable motor schemata or programs. It also codes for the sensory aspects of the mouth, including proprioception. The brain needs an elaborate "mouth map," as it needs a map for other parts of the animal's body. This takes the form of an abstract representation, not a little homunculus mouth in the animal's head. So in addition to the mouth adaptation itself, there must be a computational-machinery adaptation in the part of the brain that runs the mouth. Think of it as a set of instructions for doing things—such as seizing objects, holding onto them, letting them go, and so on. Then *that* is the adaptation that functions as a preadaptation with respect to selective cognition. The brain machinery for oral prehension is co-opted as brain machinery for selective cognition. Logically, the case is like the jawbones of the reptile being recruited as the ear bones of the mammal: a new function is assigned to an old structure, with suitable small modifications.[4] The brain circuits that underlie the functionality of the mouth are redeployed to serve the functionality of cognition, with the same abstract structure preserved. Instead of controlling only oral actions, they now also control mental actions—and these actions are adaptive, so selected for. The same program is used to run different applications, in effect. The program that was useful for controlling the dynamics of the mouth turns out to be useful for controlling the dynamics of cognition. The old brain circuits are "kicked upstairs" to perform new tasks. Thus an evolutionary step forward is taken. Our creature of the deep, with its active and questing mouth, takes the first faltering steps toward what will finally become *thought*, by developing the capacity to "hold" an object in its mind. Given ancestral preservation, our human capacity for thought therefore owes its origin to such distant goings-on, involving a fish's mouth and brain (assuming our story is on the right track). We inherited the relevant brain anatomy from this groundbreaking fish.[5]

This redeployment process can be compared to the evolution of mental imagery. First animals simply perceived the world, with no mental images. The brain contained circuits that controlled and served the outer senses. When mental imagery evolved, it did not proceed from nothing—it exploited preexisting traits, namely the perceptual systems. That is, mental imagery evolved from sensory perception: visualizing evolved from vision. The former capacity is built on the latter, with suitable modifications. But it did not do so by successively modifying the sense organs themselves, as if the eye became smaller and smaller and pushed back into the brain. There is no little homuncular eye in the brain. It did so by exploiting the preexisting brain machinery, not the outer eye. The same circuits that were involved in human vision become used in the production of human visual images (with some modifications). Thus something "outer" became something "inner." Traits of the brain are as much adaptations as traits of the body, and equally susceptible to enhancement and redeployment. Similarly, the brain circuits involved in selective cognition are borrowed from their original role in controlling the mouth. The obvious resemblance between perceptions and images suggests an identity (or at least overlap) of underlying machinery, and the formal resemblances between selective cognition and oral prehension suggest a comparable identity (or overlap) of underlying machinery. In neither case is there any objection of principle to evolution proceeding in the manner suggested. Things certainly *could* have gone this way, as a matter of logical and nomological possibility. This is an intelligible story in light of all that we know about the evolutionary process. Whether it is empirically *true* is another matter. We are speculating freely about very remote matters, about which firm data are unlikely to emerge. We certainly have not *observed* the oral brain circuits of a fish being co-opted into functioning as brain circuits for piscine selective cognition. This is a *theory*—a conjecture, a hypothesis—not an established fact. What mainly matters for our purposes is that the theory has the right methodological and explanatory properties—specifically, ancestral preservation and incremental adaptation. And what other theory meeting the requisite conditions of adequacy can be suggested? We wanted to show how thought could arise from nonthought, in a quasi-mechanistic manner, with no miracles or unexplained leaps—and this theory has the look of something that fits the bill, even if it is difficult to verify as a statement of prehistorical fact.[6] As the prehensive hands formed the core preadaptation for language, so the prehensive mouth formed the core preadaptation for thought (or one feature of it)—so we are conjecturing.

What we think of as belonging to the "mental" side of our nature evolved from something belonging to the "bodily" side (though this dualism should probably be rejected and anyway looks very artificial when considering ancient and primitive biological forms). It would be better to say: one system of brain machinery evolved from another system. The machinery for cognition evolved from the machinery for oral prehension: bits of the second machinery form components of the first machinery. Selective cognition did not suddenly get installed one day from no antecedent organic structure, as a result of a fluky mutation that miraculously had just the right properties. It *emerged* from apt and suitable preexisting traits. And once this adaptation occurred there was room for further subsequent adaptations, no doubt taking many millions of years, until we reach the biological trait called "human thought." What I have sketched here is the initial decisive step—how cognition was born (or might have been born) in its most primitive form.

I must now take up the question, postponed from earlier, of how literally we can take the prehensive theory of thought. Is it literally true that the mind "grasps" things or is this just a flaky metaphor? Does the mind "grasp" things in the same *sense* that the mouth and hand grasp things? The question is not easy to resolve. There are three possible theories of the concept of grasping: (i) the concept applies originally and literally to an action of the hand (or mouth) and only by extension and nonliterally to the mind; (ii) the concept applies originally and literally to an action of the mind and only by extension and nonliterally to the hand (or mouth); (iii) the concept applies equally and literally to both, being intrinsically neutral between the two—they are simply both species of a (unitary) genus. I suspect many people will plump for theory (i) without much hesitation. But two points should be borne on mind. One point is that if this is so then talk of the mind grasping something is pure metaphor, like calling Juliet "the Sun": but it doesn't *seem* like pure metaphor, dead or alive. How else would we literally describe what the mind does? Isn't it simply *true* that I grasp the meaning of "The cat sat on the mat"? This is not some creative figure of speech, some easily paraphrased bit of poetry. Second, it is not strictly accurate to say that the *hand* grasps; the *person* grasps *with* the hand. In the same way, it is the person who throws or holds or squeezes or signs his name—the hand doesn't do these things. To perform these actions, an agent must have certain intentions or purposes; it is not enough that his hand moves in certain ways. So the notion of grasping, even when applied to the body, is implicitly mentalistic, not purely "physical." It involves the notion of a person with purposes.[7] I would say that a robot with no inner life—no intentions and so on—does not literally *grasp* anything, even if its "hand" performs a clenching motion. It is only *as if* it grasps—just as it is only *as if* the Sun grasps the Earth in its gravitational field. Real, literal grasping, like real, literal name signing, requires suitable intentions (or at least some viable notion of purpose). Thus *I* grasp your hand (intentionally) when I take it in mine; it is not just that my hand closes vicelike around yours. It is a category mistake to suppose that the hand, considered in itself, grasps anything—as it is a category mistake to suppose that legs walk or the mouth sings. The concept of grasping is not a purely behavioristic or mechanical concept.

These reflections may tip us in the direction of theory (ii)—the concept of grasping applies originally and literally to an action of the mind, and only by extension and nonliterally to the hand or mouth—and I don't think that possibility should be dismissed out of hand. Maybe the concept of grasping applies in the first instance to the mental kind of grasping, and is then projected outward onto the hand (as the concept of *force* is projected onto inanimate bodies from its original use in application to human agency). We therefore read grasping *into* the hand (or mouth); we don't read it *off* the hand (or mouth). Don't we read grasping and gripping into the roots of a tree, because of a similarity to other prehensive structures, even though we acknowledge that roots and trees don't literally grasp or grip the earth? Maybe we *impose* the concept of grasping on the hand, viewing it as somewhat *like* the genuine mental kind of grasping. So it may be thought, but I also find this theory rather forced: it surely isn't just a metaphor that hands grasp or that we grasp with them—real honest-to-goodness grasping is going on when I close my hand around a ball, say. I do literally grip your hand when I take hold of it. So theory (ii) does not ring true, conceptually speaking.

We might then find ourselves sympathizing with theory (iii). The issue resembles the concept of the *visual* as it applies to visual perceptions and visual images. It is initially tempting to suppose that "visual" applies literally only to visual perceptions, so that "visual imagery" is really just an extended metaphorical usage. But this flies in the face of an evident phenomenological resemblance. By contrast, some may feel that visual imagination is primary, because so-called visual perception is infused with it, so that "visual" applies primarily to visual imagery. This seems hard to accept too, because visual experience without visual imagination does not seem inconceivable. Thus we reach a type-(iii) theory of the concept *visual*: there is a genus of "the visual" that is neutral between perception and imagination, and this genus has two species, both equally and literally visual. I myself favor this type of theory for the concept of the visual, and it seems attractive also for the concept of prehension.[8] There is a broad generic concept of prehension and the mind and the hand are both (literally) species of it. It is hard to articulate (in a noncircular way) the content of that generic concept without biasing it toward one or the other species of prehension, but it has to do with some abstract notion of selecting and holding (for this word too the question arises).[9] We have the idea that the mind attaches itself to an object, maybe encompasses it—and the hand does something similar. An abstract schematic notion is thus specialized into two more concrete notions. The concept of prehension, then, is equally correctly used when we speak of grasping a meaning or grasping a ball, literally in both cases.

But we do not need to insist on the literalness thesis to maintain that selective cognition originated from oral prehension, though this would certainly lend support to that theory. What matters is that we have described a possible and realistic route to the formation of selective cognition based on a preexisting trait. This is not, of course, intended as a reductive thesis—to the effect that selective cognition is nothing but oral prehension. I take it such a view is absurd. We have only claimed that oral prehension (its cerebral basis) might form a natural evolutionary springboard for cognition in this primitive sense—the seed, the germ.[10] In addition to oral prehension the organism must be an agent in some sense, and must also be sentient—without these background conditions I do not think talk of cognition is appropriate. The organism must be a perceiving agent already; only then can the mouth serve as a springboard to selective cognition. My picture of the ancestral fish is that it is already a sensing, purposive being, which has a prehensive mouth with the formal features I described; and then it makes the step to cognitive selection, with all these traits already in place. Incremental emergence over long periods by mutation and natural selection is the underlying picture, not the reduction of one trait to another. To repeat, evolution works by gradual intelligible steps, with every adaptation building on what was there before. We don't reduce the later adaptations to the earlier ones; rather, we demonstrate a step-wise procedure that respects the limits of the process. There are no "inspired" adaptations or bolt-from-the-blue mutations. There are just painfully slow transitions constrained by the existing form of the organism. For any trait of explanatory interest, we must always seek the preexisting form from which it derived as a modification. This methodological requirement imposes strict conditions on any putative evolutionary explanation. Respecting those conditions has been the main purpose of this chapter.

11   The Origin of Sentience

To be sentient is to be "able to perceive or feel things" (*OED*). It is connected to the senses: when an organism acquires senses, it becomes sentient. The senses have been around for a long time, much longer than language or even thought. The sense of touch is probably the most ancient sense. Even the lowliest organisms, such as worms, have a sense of touch—they feel things. They have sensations when they make contact with things—at any rate, many simple organisms do. There was a time on Earth at which sentience had not yet evolved, when senses had not yet appeared—the time of plants and single-celled organisms. How did sentience arise from this state of things? What was the preadaptation that formed its launching pad? We were able to come up with suggestions for the origins of language and thought—can we do the same for sentience? For that we would need to cite a trait that got *converted* into sentience by intelligible incremental steps. Let me say up front that I have no idea how that could have happened: this is one of the most profound and difficult questions facing biology, physics, and philosophy. The trouble is that any adaptation that might lead to sentience already contains it, and any that does not contain it will not lead to it. A gradualist explanation looks ruled out by the very nature of the evolved trait.

But I do not intend to debate that issue here. In fact, I am going to assume, for the sake of concreteness, that a panpsychist explanation is on the right lines—that is, sentience evolved from a kind of proto-sentience found in all matter.[1] The preexisting form for whole-organism sentience is simply small-scale sentience—sentience at the molecular and atomic level. The preadaptation for the advent of massive three-dimensional animal bodies is simply the prior existence of matter—a lot of that stuff was lying around when evolution began, and evolution tapped into it. We therefore have no problem explaining where the matter of animal bodies came from; it sprang not from nothing but from the antecedently existing material universe. Well, for the panpsychist, a lot of sentient stuff was lying around *in* matter, and the evolutionary process simply tapped into that abundant resource. Sentience did not spring from nothing either; it already existed in elementary form in all matter. Panpsychism thus has an answer to the question of *how* sentience evolved: ancestral preservation and incremental adaptation are respected. Nothing I say here will depend on the truth of panpsychism, but for expository convenience I will adopt it *pro tem*. I am more interested in the question of *why* sentience evolved: why did organisms start to sense and feel things? Why did the primitive sentience latent in matter become organized into organism-level sensory experience? It might have just smoldered harmlessly in matter for all eternity, having no adaptive potential, as it did for millions (billions!) of years before animal life evolved—so why did it acquire adaptive significance? What function does it serve? This is the same as the question: Why did the senses evolve? Why didn't life remain at the level of insentient existence? Evidently it did so for a very long time, so what changed? Or, to put it differently, why are some contemporary life forms sentient and some are not? What does sentience *do* for an organism? Why isn't the planet populated by completely insentient organisms, mindlessly going about their reproductive business, as it was for several billion years?

The theory I shall propose I call the *predator avoidance theory*. Let me expound it in a roundabout way, as follows. The theory itself is quite intuitive, but the rationale for it takes some spelling out. Before life of any kind began on Earth, when the landscape was basically just water and rock, the Sun beat down on its surface, just as it does today. The sunlight contained huge reservoirs of energy, in the form of electromagnetic radiation. This is the Stage Zero phase of life: there is a source of energy available (Earth is not shrouded in impenetrable darkness), but no life form exploits it. After a while, bacteria and elementary plant life evolved (for reasons that remain obscure), which used the Sun's radiant energy as a form of nutrient—photosynthesis began. This is the Stage One phase: naturally selected organisms living off the Sun's energy but having no sentience or accompanying senses. Stage One organisms, let it be noted, lack mobility, staying fixed in one place, unless moved by an outside force. The reason they don't move from place to place is simply that sunlight is available virtually everywhere—you don't need to travel to a particular place in order to tap into it. We can imagine a planet, call it Dearth, on which things are different: on Dearth the atmosphere is largely opaque, but holes occasionally open up in varying locations, so that sunlight reaches the surface in different places from time to time. Then plant life on Dearth will perish if it stays in one place—it needs to move to where its source of nutrition is. These sun-dependent organisms will therefore experience selective pressure to acquire the ability to move to where the sunlight is to be found (suppose a shaft of it moves steadily across the landscape at a rate of four miles an hour, with the rest of the planet in complete darkness). On planet Dearth we might well find mobile plant forms: consumers of electromagnetic radiation that follow their "food source" around. But on planet Earth the sunlight falls indifferently and abundantly pretty much everywhere, so there is no real point in moving from place to place to find it. And developing the ability to move around is a metabolically costly adaptation, so best left undone *ceteris paribus*. Foraging plants have no *raison d'être* here on Earth.[2]

Now suppose that Stage Two organisms evolve: these creatures derive their food source from Stage One organisms, that is, plants—they are vegetarian. These organisms have a fairly stationary lifestyle too, because their food source is abundant and can't run away from them (wherever there is sunlight, there are plants lazily basking in it); but they may develop some rudimentary ability to move around to deal with the fact that the local vegetation might become depleted. Think of them as slow-moving grazers and berry pickers—slothlike creatures. At this point, there is some evolutionary pressure to be economical as to size—if you move at all, as a matter of survival too much bulk is a liability (as is a cumbersome body structure). Trees can be as bulky and unwieldy as they like, because they don't have to move themselves anywhere. Stage Two organisms will have to be better designed for movement, more streamlined, less bogged down. But still rapid locomotion is not at a premium; indeed, survival can be achieved while sticking to the same general area for one's entire life. And why go to the trouble of moving, given that this is a metabolically costly activity? Why not just lazily munch away in the same comfy spot?

Now we get to Stage Three organisms: their food source is Stage Two organisms—they are carnivores. Energy is transmitted from the sun into the plants, then into the plant eaters, and finally into the eaters of the plant eaters. Stage Three organisms throw the cat among the pigeons, quite literally: they are predators. The Stage Two organisms need to avoid them. How do they do that? They need to sense when the predators are near and then move promptly away. In particular, they need to sense contact from predators, so that they can move their bodies to another place in the nick of time. Thus sentience arises from predator avoidance, as a correlate of predator-detecting senses. Costly sentience exists because being eaten by predators is costlier. This seems plausible enough so far as it goes, but there is an awkward question: why didn't the plants develop sentience too, so as to avoid *their* predators? The answer to this question is somewhat complex. In the case of plants they have a stationary lifestyle, so size isn't much of a problem (hence very big trees). But this makes them vulnerable to predation—they can't run away from the voracious plant eaters. What then is their strategy of survival? It is two-pronged: being made of (a) invulnerable parts and (b) vulnerable redundant parts. In the case of a tree the invulnerable parts are the trunk and the thick branches—no normal plant eater can destroy the plant by biting these in two, which would indeed kill it. But the leaves and buds are very vulnerable to being eaten, so how does the plant survive their destruction? It does so because the leaves are highly *redundant*—you can destroy a lot of them, maybe all, and the tree still does not die. The leaves are not *vital* parts of the tree (unlike the trunk), in the sense that damage to them leads to the death of the whole organism. Thus it doesn't matter to the tree's survival that its leaves can be easily eaten, so it is not necessary strenuously to protect against this. Accordingly, there is no need to develop senses that record when the leaf is being "attacked."

Of course, having lots of redundant parts means the organism will have trouble moving; but trees don't move, given their food source, so this problem does not arise for them.

But things are different with the Stage Two organisms: they do move. They need to move to follow their food source in case of depletion, though not very quickly, but they also need to move to avoid their predators, a good deal more rapidly. One strategy for defeating their predators would be to be composed of redundant parts, so that if a piece of the body is bitten off this has no serious effect on the survival of the organism.[3] There are plenty more legs and heads where that came from! But that strategy creates bulk and weight, and Stage Two organisms need to move, which is metabolically costly. In addition, it is inherently risky, because the predator might be powerful enough to eat the whole organism (this never happens with trees). It also involves difficult engineering problems. Once an organism has to move to survive and reproduce, it cannot afford to have redundant parts, because these will need to be carted around somehow. Thus we observe a marked difference between plants and animals: animals are far more vulnerable to damage to their parts than plants are. All the parts of an animal are pretty vital—the ears are not dispensable like leaves, say. It wouldn't matter if the vital parts were themselves *in*vulnerable—say, if they were made of unbreakable material or flesh so poisonous that it would instantly kill any predator. But for various reasons this happy setup is impossible. So the animal has vulnerable vital parts, while the plant has vulnerable nonvital parts (as well as some invulnerable vital parts, like the trunk of a tree); and this traces back to the role of movement in the organism's life, which itself reflects the nature of the food source.

An organism can in principle survive predators by a number of strategies: it can have no vulnerable parts at all, in which case it need not move to avoid predators; it can have vulnerable parts but all of them are redundant, in which case again it does not need to move to avoid predators; it can have a combination of invulnerable and redundant parts, so that again no movement is necessary; or it can have vulnerable vital parts but can move efficiently away from predators. The last is the strategy of Stage Two organisms in the presence of Stage Three organisms. But in that case the organism needs a way of detecting the presence of predators, so it needs senses, and hence sentience. Sentience arises when an organism must move to avoid predators; it is not necessary otherwise. Thus there are three necessary conditions for sentience to arise: vulnerable parts, vital parts, and a mobile lifestyle. Together these are sufficient (given the antecedent presence in matter of potential sentience).[4] Note that the first two conditions are not sufficient on their own, since there is no point in sensing predators if there is nothing you can do about it (you live by a hope-for-the-best strategy). True, if you are organism that can poison your enemies with ease, or give them a severe electric shock, then there is no need to move away from them; but this is really an example of invulnerability, and very hard to achieve given the adaptability of predators. This is why the vast majority of prey animals choose the strategy of moving away from their predators. It is not a bit surprising that predator avoidance typically involves movement and perception.

It is true that having senses would have some *prima facie* utility for Stage Two animals even before the Stage Three predators came along, because senses could serve in the search for plant food sources. The animal could see or smell a particularly delectable leaf or berry from afar and hence head right for it. But two points can be made about this possibility. The first is that the selective pressure to develop such senses would not be very intense, because the food source would not itself be a moveable feast; so acute fine-grained sensing, especially of motion, would not be required—as it is when dodging a fierce and agile predator. Second, it is not at all clear that genuine sentience would be needed to secure the advantage in question: the animal might get by with a merely insentient means of detecting the presence of food in its environment, as the most primitive organisms presumably do. It could make do with suitable photoreceptors and chemoreceptors without the full panoply of perceptual sentience. But it is another matter, again, when avoiding fast wily predators: here the organism needs the benefit of full-blown high-tech sentience—seeing, hearing, and smelling with high acuity. So I suspect that sentience only evolved on Earth once carnivores did—*that* was the urgent and nonnegotiable impetus for becoming genuinely aware of the environment. Before that it was possible to live as a sleepy and slow-moving sloth, with barely a glimmer of consciousness (maybe just a feeling of emptiness in the belly when food was needed). Given that fine-tuned perceptual sentience, like any sophisticated biological adaptation, has its own operational costs, in terms of energy consumption and upkeep of the underlying neural architecture, it will only evolve if the need for it is pressing enough—and escaping a tiger's jaws is the kind of stimulus that is especially pressing. Without the need to avoid carnivores, animals could have survived quite nicely without going to the trouble of developing the kinds of bodies and brains that make senses and sentience possible. The reason an antelope is sentient is not so as to locate the grass it eats but to avoid the predators that threaten to eat it. The senses developed in the battle against predators not in the location of a vegetarian diet. That was when the benefits of the equipment started to outweigh its costs.

It might be wondered how this theory explains the existence of sentience in the predators themselves. Predators don't need mechanisms of predator avoidance, do they? Two answers to this question suggest themselves. The first is that predators are in turn prey. This is obvious for mid-range predators, like hyenas and humans, because they are vulnerable to more powerful predators. But it is also true for top-of-the-line predators, like lions and tigers, because in youth and old age they become potential prey for lesser species. Also, they are vulnerable to competitive attack (sometimes cannibalistic attack) by members of their own species. So they need their injection of sentience too. The second reason is more interesting: once their prey develop capacities for predator avoidance, an arms race develops, because the predator animal needs to be able to detect and capture the evasive prey animal. The predator must develop sentience because the prey has. It would be different if the prey were insentient and immobile, like a tree; then the predator could get by with very minimal detection methods. But if the prey can spot you at 100 meters and make off, then you need yourself to be able to identify and track the prey over time and space. So predator sentience evolves because prey sentience evolves—predator avoidance on the part of prey is the root of sentience in the predators themselves.

The reason that animals develop advanced abilities to sense their environment, such as we observe all around us, thus traces ultimately back to the existence of predation. If there were only plants and vegetarians on planet Earth, there would be no sentience to speak of; at most there would be a very minimal kind of mechanical detection of plant food sources on the part of the plant eaters. Things only heat up sentience-wise when predators come on the scene. Then it becomes imperative to be able to detect, quickly and efficiently, what is going on around you, and take appropriate action. Advanced sentience, of the kind we see in most animals today, results from fear of other animals. The tranquil vegetarian, prey to no animal, could be to all intents and purposes unconscious—certainly not the sensory expert that populates Earth today. It is the fear of other animals that wakes him into fizzing, full-bore sentience—now he must be on his sensory toes.[5]

It is really the *lack* of sentience in plants that calls for special explanation. This is quite puzzling. Why don't they move away from their predators too? If they did, sentience would be useful to them. There must be selection pressure that favors a mobile plant (and some do move to some extent), because a species of plant that can escape its predators has a greater chance of survival than a species that cannot escape. Whole species of plants *can* be eaten into extinction, and individual plants certainly often are. Why don't plants evolve little feet to help them escape being eaten? Aren't there possible planets on which some plant species have developed locomotion? If they did, they would have a use for sentience too. The answer must presumably lie in a fundamental trade-off between the advantages of immobility and the advantages of mobility. Given the fixed solar food source, and the extra resources needed for locomotion, the plant (or its genes) makes the calculation that moving to avoid predators is just not worth the trouble. That is, the genes for its construction survive better, given its lifestyle, than in an alternative lifestyle in which locomotion plays a role. True, locomotion and sentience would be useful in the avoidance of plant eaters, but it is just too biologically costly to install the necessary machinery. Instead the plant settles for a mixture of invulnerability and redundancy. But animals that live off plants and have predators have an overwhelming need to develop senses that enable them to survive in a dangerous world.

Let me now turn to a different question: the character of specifically human sentience. Human sentience is ultimately inherited from fish sentience (the lobe fish family in particular). Genes for sentience are passed down the generations, going back to its first glimmerings. There is an ancient preexisting form of piscine sentience of which the human kind is a modification. This is as true for sentience as it is for human hands or human intestines. It is natural then to ask whether any remnants of its earliest forms survive in us to today—relics of our ancestors that no longer serve any function in us. Given that sentience first evolved in the oceans, it was adapted for that milieu; but we live on the land. Of course, many modifications of the original form of sentience have occurred—trichromatic color vision is one of them. The eye originally evolved in the water, so is there anything about the eye today that reflects this lineage? Well, one point is that sunlight is less bright in the ocean than on land, so the existing mammalian eye was originally adapted for less intense light. To come out of the murky depths into glaring sunlight would have exceeded the acceptable thresholds of ambient light—the animal would be effectively blinded. Of course, there was a process of adaptation to stronger sunlight over many millions of years, but isn't it still true that the mammalian eye is not well adapted for powerful sunlight? We can't look into the Sun; nor can other mammals—yet the Sun is a perpetual perceptual presence. Our eyes just don't respond well to glare. Not being blinded by the Sun would be an adaptive advantage. Might this be some kind of holdover from the days of ocean-going vision? Something like this is at least conceivable.

But that is more a point about optics and receptors of the eye as a bodily organ; it is not a point about the very form of our phenomenology. It seems to me difficult to identify any feature of our sensory phenomenology that reflects our piscine ancestry; we might just as well have descended from some other ancestral stock, so far as our inner sentience is concerned. On the other hand, how much does contemporary fish sentience differ from mammalian sentience when you get right down to basics? Much the same senses are possessed, so the form of the sentience will not differ dramatically. It would be different if we possessed some remnants of a sense we no longer possess, simply because our ancestors possessed that sense. Suppose we no longer had noses and never perceived smells, yet in dreaming we experienced olfactory sensations in imagistic form: this might be the result of a remnant from the old days, still with a basis in the brain but with no functionality any more. That is the kind of thing I mean by asking whether any trace of our ancestral piscine sentience persists today: the ancestral fish had a sense of smell and we experience smells in dreaming just because of that ancestry, not because we have any use for a sense of smell in our current habitat (compare the appendix). But so far as I can see, nothing like this is the case. That is not because our sentience owes nothing to the sentience of our piscine ancestors—on the contrary, ours is inherited from theirs (with modifications)—but because we haven't departed all that much from the original forms of sentience. The basic sensations of animals haven't changed dramatically from the earliest life forms—visual, auditory, olfactory, gustatory, tactile. Maybe things are different on other planets where there is conscious life. In any case, our current sentience does not seem to contain any pointless evolutionary relics from our sea-dwelling past.[6] It all seems quite adaptive. This is somewhat anomalous, given that the human body contains many obsolete relics of our distant evolutionary history (as our DNA does). Perhaps the consciousness of the fetus goes through a kind of pointless fishy phase, which is then replaced by something mammalian, as the body of the early fetus is said to display marks of our evolutionary past in the form of fishy attributes, such as gills.[7] Or perhaps it is just all too long ago and no telltale remnants survive in us from that distant period, despite the fact that we originally evolved from fish, brains and all.

Summing up our discussions so far, we may now identify three great epochs in the evolution of what we call "mind." First, after a long presentience period, in which bacteria rule the seas, basic sentience evolves, possibly because of locomotion and predator avoidance. This is then elaborated and refined into the senses with which we are now familiar. Later, cognition of the selective kind evolves, building on the prior existence of advanced oral prehension. This is then further elaborated into thought, as we know it today. Then, after a *very* long time, human language finally evolves, exploiting the hand and toolmaking and other developments. Thus we have: the Age of Sentience, the Age of Cognition, and the Age of Language. Each of these ages subdivides into further evolutionary innovations and modifications. The latter two ages were triggered and enabled principally by advances in animal prehension, involving the mouth and hand. Prehension precedes and shapes thinking and speaking. Did prehension play any role in the evolution of sentience? It might appear not, because that age was triggered by a predator-driven need for bodily senses; and indeed no prehensive preadaptation played any explicit role in the story I told about the origin of sentience. However, prehension might be seen to enter less directly into the story, in two ways. First, perception itself might be seen as prehensive in character—a mental reaching out and grasping of external things. Second, animal locomotion involves a further kind of prehensive act: the earth must be gripped somehow. The feet must grip the earth in order to propel the animal forward. This is evident when the plane of locomotion is an incline, as with climbing a tree or a steep hill. But even on flat land the feet must perform a certain amount of gripping—think of walking on a slippery surface. Claws and hooves grip the ground as the animal walks and runs. Fins can be said to grip the water in an extended sense. Wings likewise grip the air. Even in the case of snakes and worms locomotion requires a gripping action of some sort. So prehension—or proto-prehension—accompanies sentience, even if it is not a direct preadaptation for sentience. It may even be true to say that any organism that grips has sentience and vice versa. So prehension is part of the overall background of sentience, even if it is not central. Certainly the evolution of prehension marches in step with the evolution of "mind." Body and mind do not evolve as separate entities but in a kind of continuous interplay.[8]

In humans, today, sentience, thought, and language are intermingled and inextricable. But viewed diachronically we can separate these three faculties and see how they might have evolved at different times and for different reasons. The broad category of "the mind" does not do justice to this variety. An evolutionary perspective thus enables us to make distinctions that are blurred from a synchronic standpoint. Then we can formulate our explanatory questions more clearly and fruitfully. Even if my speculations are misguided, I think they at least proceed from the right assumptions.

12   The Meaning of the Grip

In this chapter I propose to engage in a phylogenetic existential psychoanalysis of the grip. My style will accordingly switch from the analytic-rigorous-scientific (well, somewhat) to the literary-expressive-humanistic. I intend, that is, to write like a "Continental" philosopher (think Sartre and Heidegger). I want to explore what the grip means *to* the gripper—its existential significance. I am speaking here of what might be called the *lived* grip—of what it feels like to be a gripping being. What are the constitutive structures of the gripping consciousness? How does the grip shape our being-in-the-world? What is our being such that in our being we grip other being? What are the phenomenological modalities of the gripping self? How is our human interiority constructed from our gripping exteriority? This is to be a no-holds-barred exercise in the "hermeneutics of the hand," from the perspective of human prehistory. It is not intended as science.

Try to imagine what it would be like to be a completely gripless conscious being: no hands, no mouth, nothing with which to seize and hold an object—one's own body, the bodies of others, the inanimate world. It isn't easy to imagine this, because we are so steeped in our prehensive powers—but try.[1] You would confront the world in a thoroughly passive manner, with no part of it ever squeezed or grasped or clasped. Presumably a disembodied self would be in this state of total griplessness. Moreover, suppose you have no memory of gripping, and are not even able to imagine gripping anything—you don't even know what it would *be like* to grip something. When you see an object passing by, you do not even have a thought that involves gripping the object. Gripping is just not part of your physical or mental being. You don't even have the *concept*. How, in that totally nonprehensive state, would you feel about the world? What would your relationship to it be like?

I suggest that your primary feeling would be one of being *cut off*. You would feel isolated, separated, sundered, and remote. You would perceive the world and think about it (we are supposing, for the sake of the thought experiment, that these are not prehensive acts), but you would experience it as at-a-distance from you, ontologically removed. The overwhelming sensation would be of a *gulf* between your conscious self and the world outside it. You would be *here* and the world would be *there*, and nothing would join the two. There would be a deep duality, an impression of complete disengagement or divorce. Phenomenologically, you and the world would belong to separate spheres. You would be a passive spectator, unable to take hold of anything. You would have no leverage on things, no purchase. You would be bereft of a vast swathe of experience, routinely enjoyed by all prehensive creatures. You would suffer a profound experiential lack, comparable to blindness or deafness or complete lack of sense experience. You would not possess a basic way of connecting to the world that is taken for granted by all prehensive organisms. You would be like a sentient jellyfish, only more so. You would be in the grip of the world, but the world would never be in your grip. You would be prehensively impotent.

What difference would adding a gripping organ like the hand make to this rather bleak and frictionless mode of existence? With the addition of the grip you would now be able to reach out and seize an object—people as well as things—cradle the object, caress it, squeeze it, mold it, hang on to it, put it down for a while, pick it up again, carry it somewhere, hand it to somebody, throw it, pinch it, turn it in your hand, get to know its shape and texture, cup it, fiddle with it, warm it, feel its solid presence. What would these actions mean to you? I suggest that the most primitive meaning of the grip is that it converts what is not mine into what is mine—it converts the separate and alien into the joined and nonalien. It takes something from *the* world and places it in *my* world. I call this operation *de-alienation*: thus the grip is a device of de-alienation. The gripless existence is an alienated existence, but the world of the gripper is a de-alienated world. The world begins in alienation—that is the primary modality of our experience—but the grip (in the human case it is mainly the hand) permits this alienation to be nullified or reduced. With the aid of the grip, objects move from *other* to *mine*. The grip is what creates the sense of *possession*—not in the legal sense, but in the existential sense. Merely seeing or hearing an object is not possessing it, but seizing and grasping an object is a far more intimate encounter, in which possession is implicated. To grasp is to acquire, to take possession of. I grasp, therefore I possess.

Consider a primeval man striding across the African savannah, with little he can call his own. He espies a nice looking round stone by the wayside. He pauses and stoops to pick it up. He likes the feel of it—the smoothness, the weight. He envisages its utility in hunting or fighting, or just throwing it for fun. It feels snug in his palm, as if it belongs. His fingers fondly encircle it.

The contact gives him a strange thrill. Happily he strides on, his trusty stone gripped tightly in his hand. The stone has become *his*—he now possesses it. Before it was pure objectivity, just a fragment of an alien and possibly threatening world; now it is his ally, his accomplice, and his friend. His being now includes *its* being. He turns it deftly in his hand, as if the two were made for each other. He is reluctant to part with it; he certainly won't let anyone take it from him. The stone is now his instrument, an agent of his survival, and an extension of his being. He feels a sense of oneness with it—of solidarity and belonging. The stone has been de-alienated. The stone has been humanized. There has been a meeting of man and object. The for-itself has made peace with the in-itself, or a fragment of it. It is no longer just *a stone*; it is now part of the human *Umwelt*.[2]

With this sense of oneness, of closeness, of possession, a range of other relationships can be formed: controlling, taming, governing, shaping, subjugating, cherishing, annexing, amassing. And it is not just the inanimate impersonal world that can be de-alienated by the grip; we can also grip parts of our own body and the bodies of others. Most of one's own body can be self-gripped, from head to toe, but some areas are hard to get to, depending on one's degree of flexibility. Some people cannot grip their own feet; hardly anyone can grip certain areas of the back. The inside of the body cannot be gripped at all—the heart, the lungs, or the liver. In the happily prehensive state we feel no alienation from our own body, though ailments can interfere with self-gripping in different ways (paralysis, obviously). Then self-alienation will likely ensue—the body will not be fully *mine*. We already feel pretty alienated from our internal organs—the ungripped parts of our bodily being. A total ban on self-gripping is likely to feel uncomfortable and gulf-inducing (consider the genitals and certain "Victorian" prohibitions). Our prehensive relationship to our own body shapes our feelings of possession with respect to it. Parts of the body are among the first things we grip, and this "auto-prehension" leaves its imprint on the psyche. I am always within reach of my own body. This body is possessed by me because I can always grip it freely.[3]

But it is in relation to other people that the de-alienating power of the grip really shows itself. The island of the solitary self is bridged by acts of other-prehension. This goes all the way from the formal handshake to the intimacies of sex. It's all grip, grip, and more grip. The more I can grip you the closer we are, and the more you are mine. The alien other approaches me, possibly unfriendly or threatening, but she extends an open hand, which I take in mine, and squeeze and hold, feeling its fleshy warmth and soft pressure—and then she is no longer quite so alien. I might not have seen a particular person for a while, so that she has grown somewhat alien to me, but I seize and shake her hand and there is *re*-de-alienation. The bridge, the link, the connection, is formed anew. "Only connect," as E. M. Forster famously advised; we might revise that to "Only prehend." Then there is the shoulder squeeze, the body hug, the prolonged inter-weaving of fingers—all the forms of interpersonal prehension. Hand-holding is the primal form of manual de-alienation: a giving and receiving, symmetrical, reciprocal, both assertive and submissive. Hand-holding says: "You are mine and I am yours." It is a possessing and a being possessed. To be bereft of any means of gripping, as in our thought experiment, is to be deprived of handholding—an existential tragedy. When gripping turns more intimate its capacity for bond-formation becomes yet more pronounced. Sex is thoroughly prehensive, is it not? What is vulgarly called a "hand job" is an act of prehensive intimacy (and many other forms of manual action could be designated with that peculiar phrase). The vagina itself is a prehensive organ, holding the (gripless!) penis in its firm embrace. The prehensive powers of the mouth may also be called upon. The whole body can be held tight and squeezed. Sexual activity is a very "grippy" matter when you think about it, and possession is its watchword. There is a continuum of prehensive closeness in human relations, and emotional bonding mirrors its gradient. The grip plays an indispensable role in the transition from alien to intimate, from stranger to beloved. Marriage itself could be seen as prehensive completion (with the finger-gripping wedding ring as a symbol of prehensive ownership).

Much more could be said about the grip and other people (as well as pet animals) but I think I have given enough of a sense of the grip-other nexus. In what Sartre calls "Being-For-Others"[4] the grip plays a vital establishing role. I certainly think this role should be recognized and cultivated (as a practical matter). When early man moved in a more social direction, forming more extensive interpersonal bonds, the role of the grip will have assumed greater importance in human life, with the handshake probably an early development. The linking of hands became common and welcomed. From gripping mainly branches the hand became an organ for gripping parts of human bodies, among other things—from brachiating to glad-handing, as it were. This is the grip as social cement.[5]

I have already discussed the grip in relation to inanimate objects. From the gripping of found objects, like sticks and stones, humans moved to systematic toolmaking and utilizing, which always involves extensive use of the grip. Thus nature becomes ever more de-alienated—rendered *ours*. The notion of *craft* embeds the grip, and craftsmanship came to define the human species in due course. In crafting a natural object we bring it into the sphere of human value ("handmade" is a term of approbation). Industry is an outgrowth of this ever-widening grip expansion. In the institution of private property (often referred to as "holdings") we find the legalized expression of primitive prehensive possession. (I shall say more in the next chapter about the hand and culture.) The human hand proceeds to denature more and more of nature, making the human world a manually constructed world-to-be-gripped. Material culture is manual culture. We now live out our days in a thoroughly de-alienated world, with our surroundings everywhere showing the imprint of the human hand. The grip has conquered nature to a considerable and unprecedented degree—so much so that there is a tendency for people to think that we *own* nature. The world is *our* world; it is radically *non*alien. Certainly there is a human aspiration to make this so—and to wish it were so. We want to remake the world in our own handy image. We want being to be being-for-us. Our will to power seeks to universalize hand dominion. If only we could mold and shape everything to our will!

In response to this anthropocentric tendency, which is both understandable and also deplorable, and to indicate the existential limits of the grip, I now want to make the point that our de-alienation of nature is always *partial*—and this too defines the human *Dasein*. We are essentially beings whose being-in-the-world is always partially alienated, despite the impressive de-alienating powers of the human hand. The project of de-alienation is inevitably incomplete—doomed as an existential imperative. The world, in fact, can never be *mine*. Why? Because all gripping, manipulating, and transforming must deal with an objective raw material that asserts its own identity and recalcitrance. The hammer may slip out of my grasp or hit my finger or fly apart, because of its nature as a material object subject to impersonal forces. The hammer is an obedient human instrument only up to a certain point—and then it becomes an alien natural object. Every tool of man is *also* an object of nature. The objective alien being of our tools cannot be gainsaid or canceled, try as we might. There is always an alien residue to every de-alienated object. In every object that is mine there is always an aspect or dimension that is *not*-mine. The grip is not omnipotent. We are always confronted by the real possibility of *re*-alienation. Tools that help us live and make us safe can also harm and even kill us. Tools may assuage our natural anxiety in the face of the world, but they are still part of that world, and so may produce the same anxiety.

Thus a profound ambivalence underlies our relationship to the world of human tools: we love them and we hate them. We play imaginatively with the idea of the perfect tool, the tool that will never let us down or harm us, but we know that this is a fiction—a piece of wishful thinking. Mold and shape as we will, the natural world will have its say. It will defy the imperious hand, and even make a mockery of it. It will bite the hand that makes it. Still, the project of de-alienation can be partially successful, and hence can alleviate the existential predicament in which we find ourselves, as angst-ridden fragile beings in a hostile and indifferent world. The hand can ward off the ever-present danger of death and destruction by means of its elaborate constructions, even if death still lurks in every object, no matter how humanly configured. What can in principle kill us can also help us live. The meaning of the tool is therefore enhanced life combined with an undercurrent of death. Think, for example, of cars and car accidents: oh, how we love our cars, but we are terrified of them too![6] The world of human equipment is a double-edged sword. The idea of the perfect tool is just a comforting myth, a fairy tale to keep the wolf at bay. No human hand, no matter how skilled and ingenious, can ever remove the fangs of nature.

Animals too can de-alienate through prehension, though their existential dreads are (apparently) less pronounced than ours. In the case of animals, the mouth is the prehensive organ of choice: what is in my mouth is *mine*. The possessiveness of a dog or cat about what it has in its jaws is palpable—not just food but objects of all kinds. Licking has a special meaning for (many) animals. Taking things into the mouth is appropriation, a seizing of assets. The animal joins itself to the world mainly via the mouth, though some animals have other prehensile organs (the tail, the trunk). When an animal sinks its teeth into something a part of the world has been assimilated into the animal's domain. Swallowing only completes the assimilation.[7]

The foregoing is one side of the meaning of the gripping action—the incorporation into the self of what was outside the self. The other side concerns attaching oneself *to* the world: self-transcendence, not self-incorporation. Not linking the world to me, but linking myself to the world. I am alluding here to what psychologists call *object relations* and *attachment theory*.[8] The isolated ego, languishing in its solitude, longs to become connected to others, to escape its solitariness. From a developmental perspective, the neonate, freshly released from the soft grip of the mother's womb, finds reattachment in the mother's breast: the breast is what is orally gripped and sucked on. Thus the breast becomes the primary object in the infant's set of object relations. The infant attaches itself, physically and emotionally, to the mother's breast, via its mouth. (This original attachment may lead to regressions in later life, according to object-relations theory.) But the infant must soon augment its object relations to include the whole mother, not just the breast part, salient and proximate though it may be (succulent too). The cessation of breastfeeding will inaugurate this new era of object relations and attachment. Then the father must be included, and later other people. Maternal deprivation occurs if the child's attachment to the mother is disrupted, because attachment relations are essential to psychological well-being and successful maturation. The happy child is the well-attached child, according to theory. But object relations can be fraught and fragile, so that there is anxiety surrounding them (the breast is always being withdrawn). Hence a fear of loss attends the child's object relations—of *dis*attachment. In these basic psychological object relations, then, we find bonding, socialization, a sense of connectedness, relief from aloneness, security, self-approbation, openness, self-transcendence, and peace of mind—all that is healthy and good. Fetishism, regression, neurosis, even insanity, can (allegedly) result from unsatisfactory object relations.

Let us summarize all this by saying that the human psyche thrives on attachment. Then the point I want to make is that attachment depends to a considerable extent on prehension—chiefly, by mouth and hand. If a child lacked prehensive capability, by accident of birth or upbringing, then object relations would be impeded and possibly completely ruptured. The breast must be gripped and sucked (or failing that some breast substitute—the "dummy"). Hence the baby is orally hyperactive, and orally centered (compare Freud's "oral phase"). Oral prehension is the original bodily mechanism of emotional attachment, according to these theories, closely followed by the hands. We reach beyond ourselves to form connections with others precisely by gripping them. No doubt some attachment can be achieved in the absence of prehension (though it is hard to obtain data about this), but given the way humans are physiologically and psychologically constituted, gripping is their primordial mode of object attachment. The infant cannot just think, "It would be desirable to become attached to my mother" and proceed to do so; he or she must go through a sequence of developmental stages in which object relations are organically and naturally established.[9]

It is easy to see that this type of psychological theorizing has its counterpart in ordinary adult human relations. I become attached to others, thus easing my essential solitariness, by means of prehensive acts of many kinds, already enumerated. I become emotionally attached by becoming physically attached. This is how I link myself to others (and to the material world). The purpose here is not possession but augmentation—expansion, enlargement. I seek to be embedded in a world that exists beyond me, particularly a social world—to become one with the universe, as people say. But I can't do it just by wishing it were so, and language has its limits in cementing bonds; the hands (and mouth) are the preferred means of securing the desired end. They are the apparatus of attachment—the hooks, the glue. If we consider early man, with his dawning self-consciousness, his limited language, and his need for others, then his hands will have played an indispensable role in forming interpersonal attachments. The hands are the *tools* of human attachment. The social being of man is a hand-mediated being (with the mouth playing its own distinctive role). And let us not forget the emotional attachments we form to purely material things, which are also experienced as self-expanding, such as tennis racquets, guitars, motorcycles, and kayak paddles. These are gripped objects, expressly so, and the grip surely mediates the felt attachment. A yearning itch in the hand to connect to these objects is a sure sign of their positive valence. There is pleasure and value in the grip.[10]

In fact, I think we can go further and defend a kind of "externalism" about our human implements—that is, these objects can become literally *parts* of the body. They are not just joined to the body, but integral to it. Thus the body externalist will claim that when I hold a tennis racquet in my hand, say, it is actually functioning as a part of my body. The fact that it is physically detachable will not disqualify it from being a part, because it is easy to imagine a "Twin Earth" on which people's fingernails (say) are detachable and yet parts of their body. The Twin Earth people, we may suppose, take their fingernails out at night for safekeeping and insert them again in the morning—yet they are otherwise just like us. What matters is the functional connection, not the ease of removal. A prosthetic hand or leg can surely be or become part of its owner's body, as much as an artificial heart or a hip-bone made of titanium. What if we discovered tomorrow that our ordinary human bones were in fact installed by clever surgeons at birth? Would we conclude that these bones are not really parts of our body after all? What if in some possible world tennis racquets grew naturally from the hands of players if they played enough tennis? Would we allow that these are parts of their body but insist that the racquets you buy in shops can never be? What if the handle grew naturally but you had the buy the head and strings? There is just no principled distinction here. The concept of a body part is very fluid and functionally defined. What counts as a part of my body is what *acts* as a part of my body.

We can thus say without undue exaggeration that the instrumental environment of humans belongs to their *extended phenotype*: our human tools are parts of our species anatomy, like birds and their nests and beavers and their dams. It would certainly be quite wrong to exclude tools from the domain of the body just because they are not located inside the skin—that would exclude teeth, hair, and nails (as well as the skin itself). No matter which way you cut the pie, the body is not in the skin! We may as well declare that the human body extends out to include the equipment we use to augment the body with which we were born. Tools are just remote prosthetic devices. We should go *externalist* about the body (the phenotype). Thus we have the *native body* and the *acquired body*, as we have innate ideas and acquired ideas (which are both still "in the mind"). The native body can be augmented with prosthetic devices that then become (acquired) parts of it. This process of augmentation extends to all human tools. With a bit of conceptual ingenuity, indeed, we could no doubt argue that a given individual's body extends out to include the bodies of others, given the right functional connections (certainly a body part of yours could become a body part of mine, as with a transplanted leg). Teamwork results in an extended *social* phenotype for the individual (the team functions as a tool by which I achieve my goals). As people have spoken of the "extended mind," so we might speak of the "extended body." From this point of view, the mind does not really extend beyond the body after all—since the body itself is already extended. The usual vague notion of the body is conventional and does not reflect the real underlying ontology. The best theoretical natural kind here takes the body to be extended into the instrumental environment. There is no natural cutoff point between eyes, fingernails, ear wax, gut bacteria, sweat, beards, prosthetic limbs, clothes, reading glasses, satchels, bicycles, cars, and tennis racquets.[11]

I have spoken in glowing terms of the grip, emphasizing its positive features. These features figure in the meaning of the grip to she who grips. But the grip also has a dark side. I shall briefly dwell on this grim aspect of gripping. There are two aspects to consider: the hostile grip of the other, and anxiety about one's own grip. We have such locutions as "under my thumb," "in her clutches," "held hostage," "grasping landlord," "iron grip of the state," and "put the squeeze on"—all of which have negative connotations. There are also unwelcome manual acts such as strangling, prodding, jabbing, slapping, punching, pinching, snatching, grabbing, groping, and so on. The hand can obviously be used to perform evil acts (though it pains me as a hand enthusiast to say so). Thus we can rightly fear the grip of the other, as well as sometimes welcoming it. Imprisonment and confinement are undesirable ways of being "held" and indeed often employ griplike devices (handcuffs, shackles). We also speak of being in the grip of addiction or insanity. The handshake, though generally positive, can easily carry a negative potential, because it can be an assertion of power or domination. An excessive squeeze of the hand can be used to threaten or instill fear, and the hand held too long in a vicelike grip is redolent of forced captivity. Thus ambivalence attends our sentiments regarding the grip. The same is true of the mouth: biting is something to be feared—and the kiss can turn quickly to the bite. Not all gripping is good, by any means. Strangling someone strikes us as especially loathsome, because it is an intimate use of the hands to literally squeeze the life out of a person. The strangler is a kind of evil ironist. The phrase "killed him with his own bare hands" is particularly chilling.

Then there is grip anxiety, grip panic. This comes in several forms: losing one's grip on something valuable, gripping forbidden or disgusting things, gripping dangerous things, and a total loss of the ability to grip. Let us focus on cases of grip fallibility. The dramatic paradigm is losing one's grip while hanging from a high place, possibly from someone else's fallibly gripping hand. The hand becomes fatigued, cramped, and unable to hold on a moment longer—and then the object or person is dropped, possibly to his or her death. Or the hand may be clumsy and unskilled, or just plain slippery. We are always dropping things or failing to grasp them properly. Because the grip is fallible it is surrounded by anxiety: is my grip good enough, strong enough, likable enough? This grip anxiety is mirrored in our anxieties about life itself: what if I lose my grip on life? The dying person is told to "Hold on!" In death we "slip away." It is as if we think of ourselves as holding life tightly in our hands, but our hands may not be up to the task. This is why dangling from a cliff seems like such a powerful metaphor for mortality: we are "hanging on to" life for all we are worth, but our grip is fallible—so life might slip away. We are also told to "seize the day" and "take life in both hands" and "grab what life has to offer." But we recognize that like all gripping, the gripping of life is fallible, sometimes futile. All gripping is inherently precarious, in the nature of the case. Whatever can be gripped can become ungripped. The grip is haunted by fear of its loss. Hence, all gripping is anxious gripping.

The value of the grip is therefore double edged, both positive and negative. It can create attachment and possession, which are good, but it can also lead to disaster and subjection, which are not so good. The attachment is shaky; possessions can be lost. The grip of the other can be a locus of dread. The hand is the soul's emissary into an alien world, but it can never in the end erase the alien substance of the world. It can at best mitigate it. The hand is a bridge, but a shaky bridge. It is at most a partial solution to our existential predicament. The self-other rupture can never be fully healed. The world can never be *my* world.

Let me end this flowery chapter with some light musings on the thumb and forefinger. In the hierarchy of human grips, that between thumb and forefinger must rank as the most exquisite. The terminal pads of the two digits fit snugly together and permit fine-tuned variations of movement, as with rolling a small round object between them. The pressure between thumb and forefinger can be precisely calibrated, and there is a high concentration of sensory innervation. Of all our hand capacities, this is probably the one with the greatest evolutionary significance, with a strong claim to making us what we are today. Is it an accident that joining the tips of these fingers produces a circle, the most perfect geometrical figure? Clearly, evolution has exquisitely tuned these delicate-yet-robust organs to work together in perfect harmony.

But what do they mean to the person who possesses them? What image do they conjure in the mind? I suggest that they signify *perfect union*. They are separate entities that can operate independently, and are different in form and movement, but they join together in beautifully calibrated acts of delicate cooperation. The two together are far greater than their sum considered separately. They resemble nothing so much as a perfect marriage. But compared to them no human marriage can match their effortless union. They work instinctively together, with no friction or conflict of interest or mutual misunderstanding. They stand as a model of the perfect partnership. They function as a unitary entity, but with separable parts: the ideal team, the flawless duet. They give us, perhaps, something to aspire to—or remind us of our own imperfect unions. They are complementary, selfless, and supremely competent. In an imperfect species, they stand out as avatars of perfection. They are one of the great natural wonders of the world.

13   A Culture of Hands

I now return to soberer territory, at least in matters of style. The advent of the bipedal gait was the critical moment in human evolution, because it liberated the hands from locomotion duties. If we were not walking on our hands, we could use them for other things. (I remind the reader of my capacious use of "we" to refer to our species and to our ancestors, going back to our arboreal ancestors. At no point was *H. sapiens* a tree-dweller, though our ancestors were. Of course, speciation takes place gradually, so that species boundaries are not really as sharp historically as they may appear today.) In our new terrestrial existence the hands were also no longer used for clinging and climbing, or much less so if we combined habitats. This led to a redeployment in the direction of tools. Tools impelled us to new psychological and physical formations, with the hand and brain evolving together. Language eventually took root, centered initially on the hands. In due course language migrated to the voice, leaving the hands free for other tasks, of which there are now a great many. As human animals refined and consolidated their advances, a further development eventually took place: the growth of culture. Thus culture was made possible by the bipedal gait, because free hands were the engine of the entire cultural process. We would accordingly expect that the origins of culture would be manifest in the aspects of current human life that we call culture. If the hands explain the Transition, they should have left their fingerprints on what they produced. Our culture should be a culture of hands.

It is noteworthy that there is no quadrupedal culture: no quadrupedal species has developed culture, despite their evolutionary success as measured along other dimensions. Not even the semiquadrupeds among our nearest primate relatives have evolved anything that compares to human culture. Their hands are too busy acting as feet, among other things. Perhaps the ideal arrangement for us would have been quadrupedal locomotion combined with a pair of free hands; then we would have had the advantages of four legs, as well as enjoying hand freedom. As it is, we must accept the drawbacks of the upright two-legged posture as the price we pay for our intelligent hands. Given the principle of ancestral constraint, the ideal six-limbed anatomy was just not an accessible trait for us: so we are stuck firmly in the line of the tetrapods. Thus we have evolved as an awkward and unsteady species, but one endowed with a rich self-created culture—instead of being a firm-footed graceful species without culture. The human creature is a tottering savant, a vertically challenged sophisticate. His brain is large, but it can easily come crashing down. In evolution everything is a trade-off.

In this chapter I shall survey the contribution of the hand to human culture. This will mainly take the form of a listing of facts. There is little that is controversial here; we just need to remind ourselves of what we already know and take the measure of that knowledge. The role of the hand in culture is plain for all to see.

Let me begin with a famous work of art, because it expresses our theme perfectly: Michelangelo's depiction of God and Adam at the Creation in the Sistine Chapel. As you will recall, this painting represents God and Adam as almost touching forefingers, with their hands at the focus of the scene, as determined by the gaze of the two central figures. The hands as elaborate pointing and gripping organs are made manifest to the viewer. Originative power is credited to God's hand, which is firmer and more potent than Adam's relatively limp hand (matched by his flaccid penis, it may be noticed). It is easy to read the picture as recording and celebrating the creative and definitive role of the hand in human existence. We come from God's originative hand and we are most ourselves in our own hands—they are the fulcrum and source of our existence as human. Spirit is infused into Adam by means of God's mighty supernatural hand, going straight into the receptive hand of the first man. Thus biblical early man is represented as a being whose being consists in his handedness. Here myth and science coincide for once. It is not a bit surprising that Michelangelo would take this view, seeing that as an artist his own being and spirit reside in the power of his hands. The creative power of the hand—it can depict even divine creation itself—would have been obvious to the artist every day of his working life. The godlike power of the artist flows through his hands, as God's power flows through his hands (I shall later comment on the role of the hands in religion generally). It is therefore very natural that Michelangelo should here depict the Creation as an affair of hands. From my point of view, this work of art is an affirmation of the originative power of the hands in forging human culture—the human spirit as it has come to exist. I think also that the artist is hinting at a minor irony about this: despite the massive and formidable muscularity of the figures of God and Adam, it is the physically much less imposing hand that carries the weight of human achievement and uniqueness. Great power can come in small packages, it seems to be saying. We take the smallness of our hands for granted, as we employ them for a thousand tasks, but they loom very large in our success as a species—they are, so to speak, our compact secret weapons. This is another irony to add to the general irony of our species success (to which I return in the final chapter). Like a seed, the hand is small compared to what it can produce.[1]

First consider visual art. From primitive cave paintings to all subsequent visual art, the hand has been the implement of creation. Early man did not paint with his foot or nose or elbow. Only the hand has the motor delicacy necessary to apply paint to surface with the requisite finesse. Perhaps man first conceived of painting when he noticed the ability of his fingers to make well-formed shapes in the sand or in mud. This was the preadaptation for pigment painting on walls: from finger painting in mud to tool painting on walls with pigment. Training the hand would be essential, but this was already in full flow with toolmaking and use. Perhaps too the representational mimicry of the hand gave early man the idea of other forms of mimicry, in which the medium of mimicry is applied *by* the hand rather than simply being the hand. Sculpture arises naturally from painting and is equally indebted to the hand. Here again early man had been molding the world with his hands for a good while in toolmaking, so the basic sculptural skills would be in place already. Sculpting is only a few evolutionary increments away from making axes and spears. It is all craftsmanship and "handiwork." Music required the manual production of instruments, and instruments are perforce played with the hands. Music is appreciated with the ears, as painting is appreciated with the eyes: but ears and eyes are impotent as creators of art—they must rely on the talents of other organs of the body. Aesthetic bliss, delivered by the senses, needs to get its hands dirty—it needs manual labor. The artist is an artisan: she works with her hands. She is also a tool-user, a descendant of early tool-users, fighting to survive. There is urgency in the artist's actions; art is not just playing around. Art is manual tool use elevated to the level of the aesthetic. Art thus has its origins in biological necessity.

Technology follows much the same pattern. Every machine needs an operator interface. A machine needs to interact with the human body in such a way that the human will can influence its actions. Thus machines always have an input part and an output part: the part where the human operator acts and the part where the machine acts. A machine needs workable controls, as well as determinate practical effects. This is as true for a hammer as it is for a spaceship. There is the "business end" of the machine and there is the "user end." The user end consists of assorted buttons, levers, wheels, pads, handles, pedals, keys, grips, and what not. Machines must be designed to do a job, but also to be accessible to the human anatomy. That is indeed the essence of a tool: an object that performs a function and can be humanly operated. Now it is true that not every machine is hand-operated, but the vast majority are; and that is because most of our machines require delicate, controlled motions and grips of which only the hands are capable. There would be no point in having the intelligence to make machines if you did not possess a bodily organ that could properly operate them. Without hands, machines would be *de trop*. Thus our technological culture is primarily a handheld culture. If we all developed some strange hand ailment that derailed our manual manipulations, our entire technological culture would grind to a sickening halt. Tools have advanced tremendously since the first primitive axes and scrapers, but in one respect they are still in the Stone Age—they still need the human hand, and that has not changed much in a long while. The design of a lever in a spaceship is not so different from the design of an axe handle. Tools and technology must be handled, and that requires a part that fits the human hand. What is amazing is how much technology has amplified the gap between what the hand can do by itself and what it can bring about by the use of hand-operated devices. But there has been hardly any advance in how tools are operated, simply because the hand has stayed pretty constant for thousands of years. If we ever develop a user interface that works directly with the brain, then tools will be differently designed; but as things are, the hand decides what designs are humanly workable. This is why the design of the hand is written all over the design of our machines, from cups to computers.

Writing obviously depends on the hand, as we humans are constituted. If language originally developed through the hands, as suggested earlier, then writing is a natural way to use language for humans. But, even with speech in command of most communication, we depend on the hand to write. The reason again is simply that no other organ of the body has the necessary dexterity and delicate gripping capability. Whales and dolphins can be credited with language but not with writing, because their fins and mouths are just not up to the task, even if other conditions were favorable for writing to develop. Writing has obviously played an enormous part in forming human culture, but it all depends on having a hand that is as agile and trainable as ours. Being intelligent enough to invent writing is not enough; you also need a bodily organ that can translate that intelligence into action, namely the hand. Fortunately the (forced) adoption of bipedalism led to such a hand. We would not be writing anything if we still lived as quadrupeds in trees. It was our dizzying descent to the ground that led ultimately to our literacy. Penmanship has replaced brachiation. If writing is the transmission of collective memory, then the hand is the agent of that transmission. It makes cumulative intergenerational knowledge possible.[2]

Play and sport make up a large part of human culture in the broad sense. The hands are inherently playful, it seems. They *want* to play. So there are a great many games involving the hand—far too many to list. It is true that not all games or sports involve the hands directly—soccer being the obvious example (but even here the hands come into play at certain points: the throw-in, goalkeeping). But the hands play an indispensable role in a great many types of game, and skill with them is crucial (remember that we must include the whole support system of the hand in our conception of the hand—wrist, arm, and so on). There are almost as many types of grip are there are sports—compare tennis and discus, archery and shot put. Eliminate the hand from games and sports and you eliminate nearly all of these activities. People evidently derive great pleasure from the playful exercise of the hands, as well as meaningful challenge. So the hands centrally gratify the desire of the human being to play—no doubt one of our deepest urges. They thus contribute to human flourishing. The hands amplify the scope of human play well beyond the play of our ape ancestors. They make us into *Homo ludens* as well as *Homo sapiens*. I like to think that in the small communities of early man there existed the opportunity for the playful use of the hand, as well as the grit and grind of toolmaking. When, I wonder, did they discover the simple game of catch? That must have cheered them up considerably. It may even have compensated partially for the loss of the arboreal habitat, with all of its opportunities for playful activity (brachiating looks delightful). Leisure activities, as well as work, center on the hand.

I have already remarked on the hands in relation to social life, so there is not much new to add. Greeting, grooming, signaling, stroking, beckoning, pointing—these are all manual acts. So are fighting, stealing, and strangling. The hand gives us the good, the bad, and the ugly. For all its virtues, the hand is also a criminal organ—a way to commit bad acts. It is difficult to be an effective thief without thieving hands. Murder invariably involves hands. The hand can save life, but it can also take life. The hand must accordingly be punished for its wicked ways—by caning, handcuffing, even amputation. The Devil finds work for idle hands, as the saying goes—devilish work. The hands are so naturally active that they turn readily to bad or imprudent acts as well as good. We thus discipline children in the proper use of their hands: "no fidgeting," "keep your hands clean," "don't pick your nose," "don't stuff your hands in your pockets," "cross your arms," "don't hit your neighbor." The hands need to be policed and sometimes punished. Nor must the hand stray where it does not belong. Some things may not be touched. Some people are declared "untouchable." Thus the hand is woven into social relations at every turn. The social fabric is threaded with digits. The creation of society, as we know it today, traces back to the liberation of the hands that followed our arboreal descent. Other social species, such as ants, rely on other mechanisms to regulate society; but in human society it is the hands that largely structure our interactions. As a social being, one of the first things a human child must learn is how to use his or her hands in public—correct hand deportment. Hand etiquette is *de rigueur*.[3]

Finally, a few words on the hands and religion. Hands play an important role in the life of Jesus of Nazareth. He works with his hands as a carpenter; he performs miracles with his hands; his hands are pierced in the Crucifixion, thus immobilizing and torturing them. In religious iconography, Jesus' hands are often depicted as slender and gentle. One does not think of his hands as sweaty or dirty or calloused. Healing (putative) typically involves "the laying on of hands." In prayer the hands are pressed symmetrically together in supplication. Priests perform ritual hand gestures and touch parishioners in stylized ways. Religious ritual is hand centered. We speak of "God's handiwork" and say that our fate "is in God's hands." Then there are stage magicians and "card sharps" with tricky hands, working their "magic." Wizards tend to perform their magical feats with hand-held wands. Magic seems to hover around the hands. They have a mystical aura in many traditions. Perhaps this traces back to those ancient evenings when humans first noticed the remarkable powers of their hands, and marveled. Maybe the member of the group with the most dexterous hands was the most revered, the most "holy." Indeed, I would say that the naturally evolved human hand is the nearest thing to magic in the known universe—though it is a perfectly natural biological organ that evolved from the fin of a fish. It is not surprising if its powers have been expressed in supernatural terms. We tend to think the most virtuous among us have beautiful hands, and the hands of the Devil are regularly depicted as hooves or claws. The hand of the Wicked Witch is always a thing of horror—bony, sharp, misshapen, discolored, terrifying. Religious, aesthetic, and moral are here intertwined in the hand, as if it condenses much in our spiritual condition. In the hand we think we glimpse our own transcendence of nature—what sets us magnificently apart. But, of course, we are as much creatures of nature as any other species, hands included; we are just natural creatures of another type, and quite continuous with what came before. The human hand, as it now exists, is a modification of a preexisting natural form, going back through countless generations to the fin of the humble lobe fish (now long extinct). But the modifications have wrought tremendous changes in efficacy and centrality. In religious imagery the hand is celebrated and elevated, though through a prism of superstition. We can appreciate the kernel of truth in this antiquated mode of thinking, though we rightly discard the supernatural trappings. The hand can be impressive without being in any way supernatural. Indeed, what is remarkable about it is precisely that it is a natural product.[4]

14 Arboreal Remnants

There is a deep irony about our success as a species, which is not I think sufficiently recognized. A few million years ago (estimates vary), our not-so-distant ancestors were expelled from their ancient habitat in the trees and driven down to the ground. This was not a case of voluntary expansion outward, like conquering heroes. It is not that they had become unquestioned masters of the trees and then added the ground to the empire of their conquests: this journey into the uncharted was not confident imperialist expansion. Rather, they could no longer live in the safety and plenty of the trees, probably because of climate change (perhaps in concert with other factors); they were forced to displace themselves and try to scratch a living from the more barren and dangerous flatlands. They were not well adapted to this new habitat. The descent (from heaven to hell?) must have been experienced as traumatic, catastrophic—more like forcible exile than exultant conquest. The mortality rate probably took a steep rise. Times must have been hard. Mere survival became paramount. The ground predators alone were a constant danger. Our ancestors, though hardy, may have been brought to the brink of extinction following the compulsory relocation. All of our hominid relatives *did* go extinct (e.g., the Neanderthals) with only *Homo sapiens* surviving into the long term. We alone had the magic touch, or the brute luck of the draw. We (our ancestors) made it through the wilderness, but just by the skin of our teeth. I picture this fraught period as one of fear, desperation, and discomfort. Our ancestors had to adapt rapidly, or die out. And adapt they did, furiously so: tools, society, and language were the major adaptations. These were in the nature of desperate expedients—like using crutches when your legs have been broken. We were like refugees in a wretched shantytown, compared to our previous life of arboreal luxury. Other ape species stayed in the trees, or around them, where they belonged, and are still happily frolicking there. Instead of picking fruit and enjoying the shelter of the canopy, existing in small intimate family groups, we had to hunt and scavenge in packs, on two legs no less, accommodating strangers, roaming far and wide, enduring many dangers, suffering the beating sun and the drenching rain, and having to find new modes of protection against predators and the elements. We just weren't *built* for the new earthbound vagabond lifestyle. And those ground predators were truly terrifying—we were totally outmatched. It was a tough life being a human in the post-arboreal wasteland. So far from being conquering heroes, our ancestors were more like bedraggled victims of cruel and peremptory nature, hanging on by a thread.

But, against all the odds, those makeshift measures—tools, society, and language—began to demonstrate an unexpected potency. After millennia of precarious existence, living on the edge, with life nasty, brutish and hellishly short, we started to assert ourselves. Our peculiar type of intelligence began to pay off in the struggle to survive. Who knew? Our tools got bigger and better. We became weaponized. We grew ever more numerous, ever less fearful. We started to gaze confidently out over the plains, instead of trembling at the edge of the forest. We spoke a richer and more useful language. Even the dominant predators started to be wary of us. This ascendancy has now reached the point where we can be described as the most successful species on the planet: secure, populous, fearless, and powerful. We have mastered nature, where once it threatened to destroy us. And herein lies the great irony: biological catastrophe led in the end to biological triumph. We snatched victory from the jaws of defeat. It is as if a wretched refugee population, driven from its homeland, managed to ascend to a position of world domination. What looked like rickety measures of last resort—crude artificial organs (i.e., tools) and hand puppetry (i.e., language)—turned out to be the most powerful adaptations ever contrived by nature! They could outdo the ferocity and speed of a lion or the venom of a snake or even the size and strength of an elephant (or mammoth: now extinct, probably as a result of rampant human predation). Amazing! We became the ultimate predators—as the end result of our initially feeble efforts to ward off the predators that confronted us on the ground. From a certain point of view, being made homeless was the best thing that ever happened to us. But it was touch-and-go, seat-of-the-pants, by no means a foregone conclusion. Our tiny twitchy restless hands are what saved us, surprising as it may seem. The great irony of human evolution, then, is that a manifest tragedy turned into a blessing in disguise: a major threat to species survival became the reason *for* survival. Stopgap measures (tools, language) became, improbably enough, awesome weapons of survival.

That is the irony that is not sufficiently appreciated. We tend to have a rather "entitled" attitude to our own existence and domination, as if it were a smooth and predictable ride to world conquest, guided by our adventurous spirit and innate intellectual gifts. Once nature had produced us, we vainly feel, the die was cast: we were destined come out on top—it was only a matter of time. But that is far too self-congratulatory and unrealistic. It certainly would not have seemed that way to our struggling ancestors out there on the African plains.[1] In fact, our survival was a matter of pure luck: the occasion of our imminent extinction became, surprisingly, the origin of our huge success. If our ancestors had stayed in the trees, courtesy of a pleasant climate, they would probably have survived as they were; but by being driven from the trees, they set the stage for the most dramatic evolutionary development in the history of the planet—us. A short-term tragedy turned into a long-term triumph.

Think of it like this. Suppose tigers lost all their teeth because of some virulent new bacterial infection that eats through them in hours and even prevents them from passing them on to their offspring. The tiger species must now survive without teeth. Their extinction looks guaranteed, and indeed much suffering ensues, with the tiger population terribly depleted. Then a few brainy tigers have the idea, admittedly desperate, of inserting small stones in the mouth to substitute as teeth. It doesn't work very well—tigers often come off badly in their hunting efforts. But it works well enough to keep the tiger species in existence, just barely (maybe the bacteria happen to produce an adhesive substance that makes the stones stick to the gums). The tigers get by, just about, relying on their shabbily designed false teeth. But then, in the fullness of time, this desperate "technology" starts to show unexpected side benefits—it encourages greater cooperation, which leads to expansions in tiger intelligence, and ultimately to language—and eventually tigers become the dominant species. Ironic, is it not? Who would have predicted it? What a turn up for the books! Catastrophe turns into triumph. Well, our tools and language are like the tiger's false teeth (remember that early human tools are just bits of chipped stone, and language is just a few crude hand gestures).

It would not be surprising if a species with this kind of past showed some inkling of its status. The shadow of its past might still fall over it. A certain fear and anxiety might attend its daily existence. For the toothless tiger, there would be a fear that the stones might run out or the natural adhesive no longer work—then what? For the treeless human, there would be a fear that our technology might be taken from us, by natural catastrophe or theft—then what? We are defenseless without our tools: our natural state (a "bare forked animal," as King Lear says) is one of total vulnerability. What if we were all struck dumb one day—or words began to seem like mere noises in the air (a virus invades the speech centers in the brain)? What if species-wide aphasia developed? We would be finished. *Finished*. We still need our post-arboreal crutches, but they could be taken from us. Thus we live with a constant background hum of anxiety—rational enough, given the facts. What would it be like to be returned to the wilderness, like our ancestors, with only our bare hands to protect and feed us, and not even language to communicate with? We wouldn't last a week. We need the safety net of our technology, and we need the linguistically mediated cooperation of others. We are constitutionally unable to fend for ourselves without these support systems, and we know it. It is as if we have a species memory of our distant parlous origins and dread being returned to that primordial terror. We know quite well what it would be like to lose our tools and our talk—those recent evolutionary acquisitions. We didn't always have them, and no natural necessity guarantees that we shall keep them. And they are all that protect us from the jaws of the hungry wolves of the world.[2]

I am speaking here of the psychological effects of our species past—of what formed us into the species we are today. In what follows I am going to speculate further about other remnants of our evolutionary history. To what extent is the present psychological condition of man shaped by his evolutionary past? This is a familiar enough question, much discussed in evolutionary psychology, but I think my speculations will be less familiar. In particular, I want to explore the psychological remnants of our arboreal past. It is often remarked that the human body bears remnants of our past as tree dwellers: our hands, arms, shoulders and trunk are those of a natural brachiator.[3] The lower body is now fairly well adapted to bipedal locomotion on the ground, but the upper body still lives partly in the trees. Thus most people (barring obesity or infirmity) can hold their body weight while suspended by their hands, and can even swing a bit. The hands can grip tight, the shoulders rotate, and the back muscles support the weight. Some people can do a few pull-ups in this brachiating position. We still have the brachiating knack, to some extent. But that is certainly not the limit of human brachiating skill: consider trapeze artists, gymnasts, and climbing children. The capacity is evidently still there in dormant form, awaiting the right training. A skilled gymnast on the high bar is a supreme brachiator; indeed, he or she arguably outperforms even the best brachiator among primates (probably the gibbon). Clearly the human body is still capable of a high level of brachiation performance. I am surprised the fitness industry has not latched onto this fact about humans and launched "brachiation training" programs, promising to exploit the natural abilities of the body and lead us back to our ancient ways.[4]

But might there not be more subtle remnants of our arboreal heritage? Might there not be odd quirks and tastes, preferences and urges, that reflect our tree-lined past?[5] Here we can only be speculative, but a number of modern human tendencies would seem to bear out the hypothesis of arboreal preservation. What we know is that our ancestors still lived in trees about four million years ago. This is not a long time in evolutionary terms. Their brains will have been adapted for tree living and this would be coded in their genes. They would be hard-wired for the arboreal life-style. When they were driven down from the trees, their brains would not have altered much, retaining the same basic psychological profile—the tree-dwelling mentality. Even as they began to evolve to conform to a terrestrial lifestyle, eventually becoming bipedal tool users and speakers, the old brain circuits would have persisted, existing alongside new cerebral machinery. Natural selection would not destroy the old circuits; it would merely add machinery for the new behaviors. Evolution always works by ancestral preservation, and in this case what is preserved is the tree living mentality—though now obsolete and largely dormant. There is every reason to believe that the old tree-dedicated brain circuits would persist to this day, folded somewhere into our cortex, idling. These circuits, however, may still be activated in some fashion in our new habitat, revealing themselves in odd aspects of our psychological life. So we should be prepared to find psychological remnants of our ancestors' arboreal brain; it would be surprising if we did not. That brain formation had existed for many millions of years before the descent, and it would not be snuffed out in a blink of evolutionary time.

Consider, then, our attitudes and emotions toward trees. We *like* trees; we like them a lot. We like to have them around, to gaze at them, to touch them, sometimes to climb them. We find them soothing, beautiful, and vaguely spiritual. We admire them, we fret about their survival; it hurts to see them cut down.

We have them in our gardens, our parks, and our city streets. We visit woods and forests. We feel that a life without trees would be a lesser life. No question, we have a thing for trees. We also love wood, the look and feel of it. We make our houses out of it, our furniture, and our musical instruments. We feel that wood is authentic and somehow good (we feel differently about metal and plastic). We are also drawn to tree imagery: pictures of trees, tree diagrams, tree analyses (logic, linguistics, biology). The form of a tree appeals to us, as if we have a cognitive module for trees. Maybe we have an innate idea of a tree (surely a gibbon does). The platonic form of a tree is embedded deep in our psyche. This is all just what you would expect given our arboreal history: our basic bodily design was forged in trees, our ancestors spent their entire lives in trees, our brains took shape surrounded by trees.[6] Thus we are *haunted* by trees. Again, I am surprised the therapy industry has not latched onto this, with "arboreal therapy" and "tree getaways" and "forest restoratives." You might actually get some peace of mind by shinning up a tree and hanging out there for a while. Supervised tree talk while stretched out on a branch or hammock might be quite therapeutic. Tree psychology could be a popular course of study (I can imagine "findings" like: "People who spend 50% of their leisure time in the presence of trees have been found to score at a significantly higher level on contentment tests than those who have no exposure to trees: see the *Journal of Arbo-Hedonics*, September 2014"). A weekend of vigorous brachiation training and intense arboreal therapy might be just what the doctor ordered—and there is a sound scientific basis for it in the science of human evolution. We might recover some of the tranquility that characterized our days as full-time denizens of the sheltering trees (living arboreal primates always look as if they are having a terrific time up there).

Consider, too, our attitudes toward heights. We have a healthy fear of them, as if schooled in the high-wire lifestyle. Living in trees, that would probably be your major fear. But we also have a fondness for heights—we seek them out. We enjoy views from high up, of open vistas, as if from the upper reaches of a tree. We build our houses in multiple stories, so that we can look out on the world from a safe and soothing height. We like our beds to be elevated, not at ground level. We enjoy balconies and turrets. Stairs appeal to us. Escalators and elevators are curiously thrilling. We endeavor to "go up in the world." We sometimes feel "down." We don't want to be "grounded." We want to "climb," professionally and socially. Onward and upward! The ground is where the dirt is; the sky is where heaven is. Hell is even depicted as *under* ground.[7] All this churning human affect is consistent with the arboreal remnant hypothesis. I do not say this is all solid scientific fact, only that it is suggestive and worth pondering, in light of the likelihood that our arboreal past still survives in our inherited brain. I doubt that elephants have these preferences—they seem quite content on the solid ground. They do not yearn for "higher things" or despise the dust and mud. Tree climbing holds no allure for them. Elephants have robustly terrestrial souls. The human soul, however, still hankers for arboreal heights, though in a muffled and sublimated way. The branches of trees are where the human heart still dwells—our ancient hearth and home. Crocodiles dream of mud and murky water, birds long for the open air, reptiles adore a patch of rock in the warming sun. We like to sit in the shade of tree and imagine climbing it.

I am suggesting, in effect, that we have a kind of tree-obsessed unconscious persisting in the crevices of our brains, as a relic of our previous mode of existence. I call this the *Darwinian*

unconscious. It is structurally similar to the Freudian unconscious, in that it is a background psychic reality that shapes
what we consciously feel and do, but it differs as to content and
etiology. It does not originate in childhood sexual trauma and
repression, but in the adaptations of our distant progenitors,
genetically transmitted to us down through the generations,
courtesy of ancestral preservation. The arboreal brain is buried inside our modern human brain, still ticking quietly away.
We do not actively repress this unconscious, as Freud would
claim, but it comes to us in disguised and modified forms, so
that some "psychoanalysis" is necessary to interpret our feelings
and actions. The *meaning* of our feelings and actions may not be
immediately apparent, because they are a confusing blend of the
old and new—as a typical town is an architectural blend of the
old and new. These are psychological remnants, not anatomical ones, and they form a complex mental stew. The Darwinian unconscious is a repository of ancient psychological traits
that have been modified and redirected, but not extinguished—
just as the body is a repository of modified ancient traits. We
thus exhibit *regressive* tendencies, as Freud also postulated; but
here our regression is to our species childhood, not our individual childhood. We are constantly harking back, as if trying to
recover a vanished past, tugged at by our evolutionary history.
Our species consciousness is thus suffused with feelings of regret
or loss or nostalgia—a kind of ingrained existential discontent.
We have never *quite* adjusted to our recently adopted terrestrial
lifestyle. We still hunger for the *Up*.[8]

It is interesting, in this connection, to consider our attitudes
toward birds—our erstwhile brothers and sisters in the trees.
Our ancestors will have spent their days in close proximity to
birds. Human beings have a distinct affection for birds. Do they

remind us of times past? Quite possibly we have an ancient cognitive module for birds, buried deep in our arboreal brain. We also admire and envy birds—they are capable of an even greater freedom from the pull of the Earth than we once enjoyed. We (our ancestors) could brachiate, which is close to flying, but they really can fly. They have not been driven down to the Earth, forced from their natural element, rendered flightless. Moreover, we resemble birds in a number of ways: we are both bipedal, with specially adapted forelimbs, both vocally indefatigable, nest building, restlessly migratory, often traveling in groups. The bird stands upright on its hind legs and it *sings*—to an enraptured audience to boot. We feel a natural affinity with birds, a sense of camaraderie. Some birds can even talk in human language. And their taste in plumage can rival our own sartorial excesses. This felt affinity is itself likely a remnant of our erstwhile tree-living days, like other forms of species recognition.[9]

But where our paths really cross is in the flightless species. These birds have descended from the heights too, and they have the vestigial wings to prove it. The wings are remnants of earlier times, like our brachiation-ready hands. Flightless birds may use their truncated wings to swim or swat off flies, repurposing an earlier adaptation, but their remnant status is obvious. The penguin and the ostrich are thus our biological brethren: they belong to the rare zoological category of the recently modified habitat-switcher. Their evolutionary history is written into their bodily form, and they have not yet cast off what is no longer adaptive. They are transitional beings—animals at the crossroads of evolution. They are neither one thing nor the other. In them we glimpse our own transformative history. It is thought that an absence of predators leads to flightless birds, and that it may occur quite abruptly. In principle, the wings could be repurposed for a new use, but in practice they often wither away. Perhaps in some species they will eventually disappear completely. Our tree-adapted hands might have withered away too, once climbing and brachiating were no longer called for; but in fact they were redeployed, and necessarily so.[10] Humans and ostriches are both newly minted, compared to most species, and hence a work in progress—as witness the perils of bipedalism and the diameter of the newborn human's head. We are treeless apes, as the ostrich is a flightless bird—a kind of zoological oxymoron or paradox. We are not so much the "naked ape" (we apparently have as many hairs on our body as the average ape, but finer[11]) as the "grounded ape." Like the ostrich, we are still feeling the repercussions of an evolutionary upheaval. We are both still in the process of adjusting. We are still finding our feet, biologically speaking. Our current body type might be strictly temporary. We have not yet reached a state of evolutionary equilibrium.

It is thought that our distant past involved not just hunting but also scavenging—indeed, for some period scavenging may have been the main type of food gathering for humans.[12] We weren't very good at killing big prey, so we went for corpses left by more formidable predators. This strikes me as poetically correct, because it is less flattering than the usual fearless hunter stereotype (since when have humans ever been fearless?). Does scavenging have any remnants in the modern human? Well, we don't tend to wander around searching for the leftovers of big cats any more, but perhaps we exhibit homologous behaviors. Some people, maybe out of necessity, do seem to like roaming around picking up what others have discarded—scrap metal, bits of wood, old furniture, seashells, and so on. This seems like a natural thing to do—a pleasant and harmless diversion (except when it leads to excessive "hoarding"). But is there any more general behavior that looks like scavenging? What do we do that involves traipsing around from place to place, picking stuff up, looking it over, deciding whether to take it home or not? Our legs may be tired at the end of the day, but we feel the effort was worthwhile. *Shopping*, of course. There is no mortal danger in shopping, you don't need to cooperate with a pack of other acquisitive individuals and share the spoils, and you might end up with something you like—food, clothes, tools, junk of all sorts. Shopping is the modern civilized form of scavenging.[13] There is something slightly shameful about it, as there is with scavenging—because you have not done the real work—but it gets the job done. It beats having to bring down the prey all by yourself. The modern department store is a scavenger's paradise, with everything within tolerable walking distance, snacks if you need them, and no problem with the weather. Here again our Darwinian unconscious is directing the proceedings, as the activity taps into ancient brain circuits honed long ago on the African plains. People will "shop till they drop" because in bygone days they needed to be determined scavengers in order to survive. Modern advertising thus appeals to the inner scavenger in all of us.

Someone may object that this is all too fanciful. Am I not just making it up as I go along? Does every form of human activity and affection have this kind of evolutionary explanation? Consider our general desire to live by expanses of water, sometimes even immersing ourselves in it: is this plausibly viewed as a remnant from our days as fish? True, our distant ancestors were fish, so should we suppose that our liking for water reflects ancestral preservation from the brains of our fish ancestors? Isn't this better explained in other ways, such as the functional utility of water and an aesthetic preference? I would agree that our interest in water is not well explained by ancestral preservation of fish characteristics in the human brain, though I would not rule out in principle possible remnants in the human psyche stemming from our fish ancestors. In this case the ancestors existed an extremely long time ago and there has been enormous change in the brains of the creatures at the origin of the ancestral line. Psychological remnants are much less likely with this kind of evolutionary distance. Also, the human affinity for living near water has a far more immediate evolutionary explanation, namely that humans need water in order to live. That is, water has great biological utility for us. It all depends on the case, which is a matter of detail and judgment. Some psychological traits may reflect ancestral preservation, while others may not—you have to examine the cases on an individual basis. I would say that our fascination with trees betrays some clear indications of remnant status, rather like our fear of snakes, but unlike our interest in water. In the case of snakes, our African ancestors would have needed a strong aversive reaction to them, given their prevalence on the African planes and in the bush. This aversion would be genetically inherited. It now exists in city-dwelling people who are in no danger from snakes at all, as a remnant from the past, not as a valuable contemporary adaptation. So, methodologically, we should not rush uncritically into remnant-style explanations of current psychological traits, but equally we should be open to the possibility of such explanations, proceeding on a case-by-case basis. I submit that the hypothesis of arboreal preservation is a plausible example of this type of explanation, as is the scavenger theory of modern gathering behavior (including shopping). At any rate, there is a *component* of these contemporary traits that looks very much like a relic from the past.

Our psychological makeup is multilayered, like geological strata over which other strata have formed. That is the Darwinian picture of the mind. As one form of psychological reality succeeds another over the millennia, according to the prevailing conditions of life, the old psychology remains lurking beneath the surface. This has been true throughout evolutionary history. In our case the ancestral psychology was forged principally in the trees and adapted to that niche, though it was modified and to some degree replaced by a psychology more suited to the ground. But it has not been totally discarded and lives on, shaping a psyche no longer ensconced among branches and leaves. Our habitat may be post-arboreal, but our minds are not. An arboreal stratum lies beneath all the subsequent accretions.[14]

15 The Future of the Hand

The existence of human intelligence poses an evolutionary puzzle—not because it is so impressive but because it is unique. Why has it not evolved in more species? By "intelligence" I here mean to encapsulate toolmaking, language, social organization, and the cognitive abilities associated with these things (call this constellation of traits *Intelligence*). It is obvious that this adaptation confers great power on its possessor—so why has no other species managed to evolve the adaptation? For adaptations like locomotion, eyesight, and a sense of smell, the utility of the trait has led to multiple occurrences across the animal kingdom—some due to inheritance from a common ancestor, some instances of convergent evolution. Isn't *Intelligence* at least as useful a trait as these? Why isn't it everywhere? This is puzzling—as if only one species among millions had ever evolved eyesight. Why, in particular, is language unique to the human species (the kind of advanced language humans possess)? If there is life on other planets, is there much *Intelligence* there? Forget the idea that humans are somehow at the predestined pinnacle of creation, toward which the universe has always been tending; our special traits are as accidental as any. It is not that the universe has been working toward producing *Intelligence* all along and picked us alone as its proud bearer. It might never have evolved at all.

The answer must be that *Intelligence* is simply not an accessible trait for other species. But why is that? No doubt it would benefit them greatly if they had it, but there is no platform in their current condition (nor in their past condition) from which it can be naturally reached. It exists in logical adaptive space for many creatures, but in empirical reality it has only evolved once on planet Earth. But *why* do other species lack the necessary preadaptive platform? What did we have prior to our development of *Intelligence* that they don't have? What did early man harbor in his biological makeup that made *Intelligence* possible for him?[1]

It would be a mistake to summarize the thesis of this book by saying that he had a tree-adapted hand. That is obviously not sufficient, because our primate relatives also have tree-adapted hands (and have had for a long time), but they lack *Intelligence*. Rather, it was the forced descent of this hand from the trees that was the decisive factor: the liberation of the hand from locomotion duties, the subsequent development of tools, and so on. It was the modifications to the arboreal hand, occasioned by the descent, which made the difference (the elongated thumb, in particular). The Transition occurred because the hand had to adapt to terrestrial life, along with the brain. Fortunately, it was capable of the kinds of transformations required, these being essentially incremental (mainly the gradual lengthening of the thumb). Ground-dwelling species in general don't have hands, only arboreal species do; so terrestrial animals never developed *Intelligence*—as persistently arboreal species never do either. It was the combination of possessing an arboreal hand *and* being relocated to a terrestrial setting that set the stage for *Intelligence*—the odd confluence of these circumstances. It was the *mismatch* between body and niche that provided the essential preadaptation—the fish-out-of-water aspect of the whole thing. The route to *Intelligence* went through the freak accident of an arboreal hand no longer doing arboreal work. It went through the fact that the hand had been made pretty much *redundant*—and hence available for other types of labor. It is as if the world went completely dark, so that eyes were no longer functional, but accidentally nature found a new unrelated use for eyes—as it might be, using them as flotation devices, or making sounds with the eyelids. Imagine if such a peculiar train of events led to the world domination of the species that discovered this new use for eyes: then that is a bit like what happened with us. The explanation of our species success resides in the unlikely redeployment of an organ stripped of its original function, namely tree clinging and brachiation. The success arises, fundamentally, from the disenfranchisement of the hand, not from its continued use in its natural environment. On the ground, natural selection operated much more harshly than it had in the trees, and the result was a rapid change of phenotype that led ultimately to *Intelligence*.

Now we can see just how improbable the evolution of *Intelligence* was. It required a remarkable confluence of unrelated circumstances. There is nothing natural or inevitable or even vaguely foreseeable about it. It is not that any hand, in the fullness of time, will naturally power the growth of *Intelligence*. It just so happened that our tree-adapted hand, forced into the alien terrestrial niche, and after some minor anatomical tinkering, proved to contain the preadaptations necessary for tool use and language—though these traits were certainly not written into its original design. This is surely a virtually unrepeatable set of circumstances. Throughout the course of evolution all species have got by without the aid of *Intelligence*; we alone evolved the trait. And the reason we did so is not the kind of thing that naturally repeats itself. Unless something analogous has occurred on other life-bearing planets, which seems very improbable, we may expect to find a distinct lack of *Intelligence* distributed throughout the universe. The circumstances under which we acquired *Intelligence* are just so improbable in themselves that they are unlikely to be repeated. Our biological line would never have developed the trait if it weren't for the descent from the trees, combined with the antecedent structure of the primate hand. That we did develop it was just a monstrous piece of blind good luck (based on bad luck). Thus we see the answer to our puzzle: *Intelligence* is unique because the preconditions that led to it are virtually unrepeatable.[2] Indeed, our theory of the emergence of *Intelligence* is confirmed by the very fact of its uniqueness. Any theory of that emergence needs to be able to account for its uniqueness. How often has evolution thrown up a case of dispossessed hands? Unless this odd scenario is repeated on other planets—and why should it be?—humans may be the only species ever in the universe to evolve *Intelligence*.

You might think that *Intelligence* could arise from some other cause—it might be able to arise under a variety of conditions. In principle that is possible, but it is striking that it has not in fact ever arisen from any other cause on Earth. Maybe a species on another planet has evolved it as an answer to some other pressing need and on the basis of a different anatomical endowment—say, in order to negotiate labyrinthine pathways and mediated by extremely complex eyes. This species, we may suppose, has no hands and has never been ejected from its ancestral habitat, yet it evolved *Intelligence* nonetheless. Maybe—but on planet Earth the lines of evolutionary development seem to preclude any route to *Intelligence* except the route followed by humans.[3] No birds, reptiles, or even other primates, have found a way to acquire the trait, despite its utility; and we have it only because of an oddity of contingent evolutionary history. Our status as *Homo sapiens* rests on a freak historical event, a wildly unlikely series of accidents. In no way were we ever destined to become *Intelligent*. It was entirely possible that the complete history of life on Earth, from its first glimmerings to its eventual total extinction, extending over billions of years, should have *never* produced a species with *Intelligence*. As things are, it has done so only once, despite eons of time in which to stumble on the adaptation, and despite the existence of millions of distinct species. So this adaptation is quite unlike other adaptations such as locomotion and eyesight, which are quite predictable from the general conditions of life on Earth—which is why they are so widespread. *Intelligence* deserves to be called a "singularity."

As Darwin remarked, we are clever because we are feeble.[4] We might even say that we are unnaturally clever because we are naturally feeble. Our cleverness is a kind of compensation for a lack of natural power. Without brawn, you must develop brain. Cleverness is our adaptation for *not* being well adapted— to ground living, life under (not in) the elms. But cleverness—a big brain—is pointless unless it is harnessed to a body capable of making use of it. The hands are the primary bodily means whereby we manifest our cleverness. No other organ comes close to the hands as an instrument of intelligent action. The hands are capable of expressing every nuance of our intelligence. They have, as I have repeatedly remarked, served us well. They saved us from extinction and gave us comfortable homes and personal security—as well as the delights of culture. But will they continue to occupy center stage in human existence? What is their evolutionary future?

It is hard to see how the hands could become obsolete, given their role in the operation of technology. Manual skills of traditional kinds may fall away, as work becomes ever more automated. Penmanship has surely had its day. But surely the hands will be necessary for the conduct of daily life far into the future (the legs will be obsolete sooner). I can envisage just one way in which the hand may become a thing of the past, withering into a useless appendage. That is by the invention of direct neural manipulation: issuing orders straight from the brain into the tool we wish to use. This would require a device, attached to the head, which translates neural signals into electromagnetic energy that is then transmitted to an input module on the instrument we wish to manipulate—willing things to move, in effect. Thus we could drive our cars without laying a hand on the wheel, engage in hands-free cooking, and knock in a nail without holding a hammer.

But this is a narrow utilitarian view of the function of the human hand: we would still want to use it for purposes of play and social relations, would we not? The only threat to this that I can envisage is some new form of zealotry that regards the hand as offspring of the Devil and prohibits its employment. Given that the brain transducer has made the hand obsolete for work, this anti-hand zealotry might succeed in banning its use elsewhere. I can imagine a futuristic science fiction story, of dystopian tendency, being written around the theme: *Doom of the Hand*. It might prove an instructive tale. But if such a thing ever came to pass in reality, it would surely be a great tragedy for the human race. It would, I suspect, cause a sickness of the soul far greater than any technological or social change yet inflicted on us. It would be the end of the human race as we know it.[5] For that reason I doubt such a thing could ever come to pass. The hand will live on.

Although I don't think the hand is an endangered species, I do think we should regard it with greater reverence than we are apt to. We should give it the credit it deserves. This is why (with tongue in cheek) I advocate forming a "Cult of the Hand" to raise its profile. The cult will celebrate the hand in all its glory, acknowledge its achievements, and seek to improve hand awareness and hand health. This book might then be viewed as a text to motivate and underpin the formation of such a cult. I therefore cordially invite my readers to join the Cult of the Hand.

Notes

1 Origins of Humanity

1. A standard textbook is McKee, Poirier, and McGraw, *Understanding Human Evolution*, 5th ed. In what follows I shall take this material as a given.

2. I shall often refer to the human species as "man," which is traditional, but no exclusion of women is (of course) intended; I use this form simply for stylistic reasons, and compensate for it in other ways in the text. I also use it to refer to ancestors of ours that are not yet strictly of our species.

3. Bell's book is a so-called Bridgewater Treatise and is intended to work as an instance of the argument from design, based specifically on the hand. We can appreciate the force of the premises without acceding to the conclusion. Bell is writing pre-Darwin and so fails to see any alternative to divine design. In fact, as Darwin later argued, the hand is an excellent argument *against* Creationism, because of the affinity of the human and ape hand.

4. It is actually more Darwinian to reverse Darwin's formulation: it is not that the hand is adapted to act in obedience to man's will, as if the will existed independently of the hand; rather, the will is adapted to the hand—a product (partly) of the powers of the hand. We have the will we do largely because we have the hand we do.

5. Thus Steven Pinker says, in *How the Mind Works*: "Hands are levers of influence on the world that make intelligence worth having. Precision hands and precision intelligence co-evolved in the human lineage, and the fossil record shows that hands led the way" (194).

6. I use the word "prehensive" frequently in this book, though it does not appear in the *Concise Oxford English Dictionary (OED)*, to mean "relating to prehension." I sometimes use the familiar "prehensile," but the word has a somewhat narrower connotation than "prehensive." In the *OED* "prehensile" is defined as: "(Chiefly of an animal's limb or tail) capable of grasping." It would be odd to speak of prehensile mental acts, but speaking of prehensive mental acts occasions no linguistic recoil.

7. Apart from the works already cited in the text, I recommend Johanson and Edgar, *From Lucy to Language*; Gibson and Ingold, *Tools, Language, and Cognition in Human Evolution*; and d'Errico and Backwell, *From Tools to Symbols: From Early Hominids to Modern Humans*. For evolutionary background, see anything by Richard Dawkins, but especially *The Ancestor's Tale*. I should also mention Raymond Tallis's *The Hand: A Philosophical Inquiry into Human Being*, which contains some interesting speculations and historical background, though it is occasionally a little oracular for my taste. Recently published is the anthology *The Hand, an Organ of the Mind*, edited by Zdravko Radman, which contains a number of excellent essays very congenial to the approach favored here.

8. Partly I am motivated by the sense that philosophy of language in the analytical tradition has not been naturalistic enough—it has not located its subject matter in the broader biological world (this is also a theme of John Searle's). But the naturalism I have in mind is not the doctrinaire reductionism one often encounters but (to borrow a phrase of Galen Strawson's) *real naturalism*, i.e., relating language to the real biological sciences, not some philosopher's preconceived idea of what counts as scientifically kosher.

9. There were, of course, many hominid species, now all extinct, that preceded the strictly human line—the Australopithecenes, *Homo erectus*, *Homo habilis*, the Neanderthals, et al. Our particular hominid species evolved through several stages of speciation, linking us to more apelike

creatures—there was not a big jump from ape to human. I will simplify all of this by speaking of "our" evolution from apelike creatures, though this use does not denote humans alone. Indeed, much of what I have to say could be applied to the intermediate hominid species, on the assumption that they also achieved something akin to the Transition (for example, if Neanderthals developed language). So I am not really concerned with the human species as it exists in recorded history; I am concerned with the much earlier stages in which language, rational thought, and primitive technology arose—probably before agriculture.

2   Two Evolutionary Principles

1. This was known even before Darwin and provides a fine example of exaptation, whereby a structure acquires a new function by natural selection. The ossicles of the mammalian inner ear have their origin in bones used to operate the reptilian jaw, which became detached from the jaw and better suited to detecting high-pitched frequencies. Many intermediate forms of these migrating bones can be found in the fossil record.

2. The genes produce copies of themselves, as moving bodies "copy" their state of motion, until acted on by an outside force—natural selection or external influence. Animals don't change unless they have to, as moving bodies don't change their state of motion unless acted upon. That is, constancy is the default state of things.

3. This is an important difference between biological evolution and cultural evolution, which can proceed by abrupt, creative leaps forward. Human creativity is not inherently conservative and gradual. Modeling biological evolution on cultural evolution produces misplaced tolerance for saltationist explanation. This is precisely where Creationism goes wrong, among other ways.

4. For language, saltation looks especially tempting, because of the absence of intermediate forms in Earth's existing fauna and the fossil record. But we have to resist this temptation, on pain of postulating miracles. Language must have gone through many evolutionary stages

once the earliest forms were off the ground, just as limbs, eyes, and teeth did.

3 Human Prehistory

1. I say "virtually" to allow for some primitive tool use, as with contemporary apes that use stones to crack nuts and stems to fish for termites. I see no radical discontinuity between such tool use and human tool use, just a (big!) difference of degree. Nor do I see any radical discontinuity between primitive tool use and earlier forms of behavior, such as swinging from branches: both involve purposefully utilizing the resources of the environment. I discuss tool use later.

2. This use of "we" will not seem stretched if we remember that species boundaries are quite vague, with many intermediate forms. I am, of course, not suggesting that the species *Homo sapiens* ever lived in trees (except for contemporary humans who choose to live in tree houses). Members of our *extended* species were tree dwellers—that distinctive zoological form. We might indeed have been able to interbreed with our arboreal ancestors had we existed then. In any case, the point is that our primate ancestors lived mainly in trees.

3. The liberation was not all or nothing, because the prehuman arboreal hand would be free in the sitting position, and apes today use their hands for tool use in that position. But the bipedal posture allowed for far greater liberation, so that tools could be used while walking—hence all the time. Also, the adaptation to other uses of the hand would not interfere with its use in locomotion, as it would for a quadruped. The bipedal ape could therefore dedicate its hands to nonlocomotion tasks and adapt them accordingly. Hands with a dual use (locomotion and manipulation) will be more limited in one of these uses than manual specialists.

4. If our hands had failed us on the ground, proving unsuitable for tool use and verbal signing (see chap. 8), we would probably have gone the way of the dodo, because the rest of the human body is not well adapted for competitive survival on the ground, compared to other ground-liv-

ing mammals. In particular, if the thumb had lacked rotational mobility we would probably have been over with as a species. Even today, if we all lost the use of our hands it is doubtful that we could survive as a species. Other living arboreal primates would surely go extinct if forced down from the trees to which they are adapted.

5. Napier says the following of post-arboreal life: "Food would be less easy to come by than in the forest; predators would have abounded and escape would no longer be a matter of fleeing at breakneck speed through the security of treetops high above the ground. If, in addition to these hazards, our early ancestors were in the process of adapting their gait from quadrupedalism to bipedalism, then it is difficult to see how they could have survived the transition" (*The Roots of Mankind*, 172–173). Yet survive we did—the question is how.

4 Characteristics of the Human Hand

1. A large thumb is an impediment to a habitual brachiator because it is apt to interfere with the hook grip used to grasp branches. Only when no longer employed for this function can the thumb comfortably lengthen; and this proved our salvation on the ground.

2. The lobe-finned fish, as opposed to the ray-finned fish, is now extinct for the most part (the lungfish still exists). Its special fins enabled it to propel itself on dry land, thus leading to an amphibious form. The bones in the fins of these fish resemble the leg bones of contemporary tetrapods.

3. See Napier, *Hands*, chap. 2, for a detailed description of the structure of the hand. Since the hand is "metabolically expensive," both in itself and with respect to its brain machinery, we can only suppose that its remarkable capacities have strong evolutionary significance—they are not mere ornaments, but vital organs. The fundamental abilities displayed by a trained pianist are part of our biological inheritance, not mere cultural luxuries. The evolutionary importance of the hand can be measured by its acrobatic powers: we *need* good hands, strong, agile, and sensitive.

4. I don't mean that the hand literally remembers—its owner does. I am speaking of the brain resources dedicated to memories we express by the hand. Still, the hand can often seem like a source of agency in its own right, so the anthropomorphic metaphor is apt.

5. It would be interesting to conduct cross-cultural studies of possible "hand universals" and other evidence for innateness, as well map out the child's sequence of maturing hand skills. One can envisage a "poverty of the stimulus" argument here, as well as lack of explicit instruction in hand skills combined with a uniform acquisition of such skills. The functional abilities of the hand are likely to be as innate as its anatomical structure, as we would expect for the hands of other primate species (as well as feet, fins, claws, etc.).

6. I doubt that we can make such a strong claim for any other part of the human body, say, the feet. In humans the body is, so to speak, always cocked for hand action, which is pretty constant. Nor can I think of another species in which so much of the body is so patently geared to serving one part of it, though I suppose the jaws of a predator come close. The hands are where the action is and the body knows it.

7. But not the best nose or ears or eyes or jaws or legs—here we are outclassed by many other species. We are quite mediocre with respect to these endowments, but when it comes to the hands we are in a class of our own, the clear gold medalist.

5   Hands and Tools

1. See McKee and Poirier, *Understanding Human Evolution*, and Gibson and Ingold, *Tools, Language, and Cognition in Human Evolution*.

2. It is hard to think of another species that had to make such a large adjustment and survived; in this we see that man is *adaptable*. Since he was poorly adapted to savannah living, he had to compensate with adaptability. Thus man is a maladapted adaptable in his biological being. His major adaptation is adaptability, which is a variable, not a constant. In tool use we see man's flexibility, not his naturally given expertise. Man is the first and only evolutionary nonspecialist. The hand is the organ through which his adaptability is expressed.

3. The point then is not that human tool use is qualitatively unique—it does have instances in other species—but that the important distinction is between animals with a certain cognitive configuration and animals without it. It is the tool-using mentality that is crucial—this is what has enabled us to become what we are today, not certain kinds of external manipulation as such.

4. We might even say it was the introduction of imagination, or its large expansion, that enabled humans to cope with their new world by the use of tools. It was physical tools *plus* imaginative thought. This mental faculty must have had its precursor in the minds of our ancestors, but it was powerfully selected for in our new terrestrial lifestyle. When things are not ideal as they actually are, it is necessary to imagine alternatives and bring them about: this is what we are good at.

5. Here as elsewhere we have a continuum of abilities, not a sharp cutoff point. We can say that human tool use differs from ape tool use *quantitatively*, not *qualitatively*, but some differences of quantity can be enormous. Apes will never invent and build the automobile, but they do have a solid grasp of the mechanics of fishing rods. What they mainly seem to lack, in comparison to us, is the ability to invent complex multipart tools that require assembly and maintenance.

6. Of course, I am speculating, as I promised I would. Tool cognition itself leaves no trace in the archaeological record, unlike tools themselves, so I am inferring early man's psychology from what is left physically. Aside from behaviorist prejudice, I think there is little doubt that the range and type of tools found in archaeological sites provide strong evidence for the kind of instrumental cognition I am describing. I suggest that we should not be shy about forming psychological hypotheses based on physical finds—after all, this is really just a special case of knowledge of other minds. I would say the same about the cognitive capacities underlying the tool use of contemporary apes.

7. Many ancient tools appear designed for cutting and scraping carcasses—so that these tools were developed in the service of *butchery*. Early man was a butcher by trade, before the division of labor set in. One wonders how this has shaped man's sensibilities: he was a butcher

before he was a baker or a candlestick maker. Any squeamishness he may have felt had to be suppressed.

8. This early "tool-philia" has reached its apotheosis in modern man, whose outlook is so thoroughly saturated with instrumental thinking and tool mania that we find it hard to conceive the world in any other way. Nature for us consists in raw materials for tools. The passion for consumer goods is in the same vein. Perhaps some of the driven quality of our tool acquisitiveness derives from the urgency of tool acquisition in early human societies.

9. Just as the hands make tools, so tools make the hands—both phylogenetically and ontogenetically. Our hands are formed by the tools we use, of necessity. We have the hands we have because of the tools we use. This is not true of other parts of the body, though they may operate tools to a limited extent; nor is it true of other species—apes do not have tool-created hands.

10. We are still very impressed with feats of the hand—hand amazement is commonplace. Thus we marvel at concert pianists and guitar "shredders" who move their fingers at lightning speed, or jugglers and prestidigitators.

11. Many fossilized skulls show the imprint of big cats' teeth, as the slaughtered human is dragged off for leisurely consumption. The production of fire may have made all the difference in fending off predators, which is also an achievement of the hands. Fire has been one of our most effective tools. Fire may also have been an important shaper of the human hand, as we evolved hands that could manipulate fire without getting burned.

12. I know of nothing in the philosophical literature on language that makes any connection between language and tool use—apart from Wittgenstein's later comments on words as analogous to tools. The governing image has been that of a mathematical structure or an act of speech, not practical tool using. But anthropologists routinely put the two together, as we will see in the next section. We need to see language as more grounded in the body than we are accustomed to.

6 Hands and Language

1. Two useful collections of papers on this are Gibson and Ingold, *Tools, Language, and Cognition in Human Evolution*, and d'Errico and Backwell, *From Tools to Symbols*. It is notable that there is no philosophical contribution in these books; I am trying to fill the gap.

2. The question then is how inner reference and predication become externalized, embodied in a public communicative system. They need an *apt* basis in something external—that is, something with the right intrinsic properties. It needs to be perceptible, voluntary, and highly versatile—the eyelids or knees would not do.

3. The gestural theory of primordial language has been around a long time. For a recent endorsement see Armstrong, Stokoe, and Wilcox, *Gesture and the Nature of Language*; for a more popular discussion, see Wilson, *The Hand*, esp. chap. 10.

4. Probably it was a hybrid system combining vocal emissions and gestures, with gestures dominating. The human vocal tract developed relatively late in the hominid line: apes don't have such a tract and prefer gesture as a means of communication.

5. See Armstrong, Stokoe, and Wilcox, *Gesture and the Nature of Language*. A study of gesture is McNeill, *Hand and Mind: What Gestures Reveal about Thought*.

6. This kind of bimanual action would characteristically occur during toolmaking. It is essentially a triadic structure: right hand, left hand, and tool. It is important that two actions occur simultaneously: one of gripping, another of acting on the gripped object. Even one-handed exercises display this duality: first the spear or axe is gripped, then it is hurled or brought down. Similarly, a cup is gripped and brought to the lips—one action presupposes the other. Since many, though not all, hand actions display this duality—we could call it the "grip-and-do" pattern—we can see how the duality might underpin the duality of reference and predication, with gripping as referring and doing as predicating. We grip in order to do, as we refer in order to predicate.

7. Remember that we are seeking merely a precursor structure, not a conceptual reduction. Gripping by the hand must be converted to reference by modification; it is not already reference. Supplementary factors must be brought in. At best, gripping is a kind of proto-reference—scaffolding, not the completed building. In the next chapter I discuss in more detail how the transition might be effected. For now I am just trying to isolate a precursor trait that might be upgraded to reference of a simple kind. It is not, then, that referring *is* seizing; rather, it derives from seizing, when supplemented in various ways. Referring *evolves* from prehending, suitably supplemented.

8. I am alluding here to the Wittgenstein of the *Tractatus*, with his "picture theory" of meaning. As a man trained in engineering, Wittgenstein would have been very familiar with handheld tools, and with the congruence between tool and grip that they entail. The grip is isomorphic with the tool, yet different in kind from the tool. Could this have subliminally influenced his thinking?

9. Mimicry is actually very common in nature, from the octopus and cuttlefish to human impersonators. Some species even mimic other species, usually for protection. Primates in general are very prone to imitation. The hands make the perfect vehicle for more abstract kinds of mimicry: instead of copying with the whole body, the digits can be used to symbolize objects. All mimicry involves representation, but hand mimicry takes it to a new level. Thus we see a continuum from simple duplicative mimicry to more structural forms of mimicry.

10. It is important to be open to the possibility that the thing we call "language" and assume to be unitary is really a patching together of disparate traits antecedently existing in the human line. There was not a single preadaptation that led to language but a converging collection of prior traits. Language has many seeds—though the hand is the soil in which they all grow. In its origins, language could be quite *messy*. Always remember, in evolution nothing is preconceived—things just happen for multifarious reasons. It is always coincidence and confluence, chance and the lucky break. Language was not intelligently designed but randomly hit upon; and its roots are likely to be multiple and unconnected.

11. If there is a *language of thought*, as many suppose, then language was already installed in the head before it found a home in the hands. It might have stayed in the head, if it were not for certain historical contingencies, such as enforced terrestrial living and an adaptable hand. In any case, linguistic structures were already exemplified in the human brain, if Mentalese exists; the problem was how to get them out into the observable world beyond the brain.

12. It may well be that without the prior adaptation of the hand in tool use and construction it would not have been capable of acting as a means of linguistic communication. First, tool use led to brain expansion and hence greater intelligence, which was required for language to originate. Second, tool use led to enhanced motor powers in the hand, especially the fine-tuning of the fingers. Thus man could not have skipped the tool-use stage and progressed straight to language; he had to do his apprenticeship in the tool shop. Without millions of years of natural selection for tool use humans might never have reached the anatomical and cognitive threshold for language to develop. We might have been stuck at the symbolic level of the smartest apes. Here we see that evolution is a complex, unpredictable, opportunistic process, in which it is only possible to achieve one advance by going through another. It was the tool-hand nexus, developed over countless generations, which laid the foundation for language to develop, which itself was a complex multilayered process extending over millennia. Language did not descend from the sky fully formed one bright morning.

7   Ostension and Prehension

1. The cupping grip is intermediate between the occlusive closed fist and the flat open hand—it displays while stabilizing the held object. But for a live insect or small animal, or even for a piece of fruit while on the move, something more forceful is needed—hence a tightening of the hand around the object. Here the hand acts like forceps.

2. The notorious blindness of even quite intelligent animals to the pointing gesture might then be put down to a deficiency of imagination. Apparently chimps understand pointing, so they must be capable

of imagining the extended index finger, according to the theory proposed here. Without imagination, reference is impossible, on this view. In other words, meaning and imagining are intertwined—a not unfamiliar thought.

3. Note that the preadaptations for pointing form a suite of preexisting traits that are united in the new adaptation. Thus we have imagination combined with prehensive action; but also teleological thinking, tool use, and social cooperation. Pointing may look simple but it rests on a constellation of coordinated factors.

4. If we wanted to refer to things and events in the past or future, we would of course need time-traveling arms—a tall order. Alternatively, we could resort to deferred ostension, whereby a present entity standing in a suitable relation to something in the past or future is exploited—as with pointing to a man in the past by touching his present footprint.

5. There is a whole subject here that will remain unexplored—imaginary gripping and touching. We certainly do imagine gripping and being gripped, sometimes compulsively; but we also imagine forms of gripping that are not exemplified in the real world—satanic or godlike gripping, say. In addition, we imaginatively extend the notion of gripping to things that do not literally grip—as when we speak of the "grip of gravity." We seem quite expert in imagining variations and transformations of the empirically perceived grip, so our understanding of pointing might well incorporate such imaginings. Maybe early man had a vivid mental image of an extended index finger when one of his fellows pointed at a distant object, but now this has faded to a kind of implicit imagining.

6. All these actions have a kind of ostensive intentionality built into them—an object of reference. The air kiss has a certain target, as does the fist pump, as does the wave or beckon. This is possible only because of a certain cognitive background that surrounds the physical action, possessed by both agent and audience. One needs to have a certain object "in mind." By themselves, the physical actions are just meaningless movements in space. I am classifying ostensive pointing along with these other object-directed intentional acts.

7. What is called the *causal theory* of reference tends to model referring on perceiving, because perceiving is a causal process, whereby the external object impinges on the senses. By contrast, the haptic theory of reference emphasizes the active nature of touching by means of the hand, not just the passive receiving of sensations. This fits with the fact that referring is a type of action. I take the haptic theory to be of a piece with the "real naturalism" that I advocate, as opposed to the usual reductionist efforts.

8 From Signs to Speech

1. See Armstrong, Stokoe and Wilcox, *Gesture and the Nature of Language*, esp. chaps. 1–3.

2. Darwin made this point in *The Descent of Man*, 32–33.

3. The hands are more honest and forthright than the voice—more "transparent." The whispering voice can be more devious and cunning. I would not be surprised if the desire to whisper had a greater role in the transition to vocal language than we might naïvely suppose.

4. Apparently there are close connections between cortex devoted to the voice and cortex devoted to the hand, so that aphasia and apraxia are often correlated: see William Calvin, "The Unitary Hypothesis: A Common Neural Circuitry for Novel Manipulations, Language, Plan-Ahead, and Throwing?" in Gibson and Inghold, *Tools, Language, and Cognition in Human Evolution*.

5. Language leaves no fossil traces, so the time of its origin is moot. However, it is quite wrong to assume that human language began when the specialized human larynx formed, which is a fairly recent event (the Upper Paleolithic, about 50,000 years ago). For language may have existed in gestural form for hundreds of thousands of years before this anatomical development. My own speculation, for what it is worth, is that language goes back at least as far as the Neanderthals. The more we learn about the Neanderthals the more sophisticated they appear, contrary to the popular myth; and given their social nature and capacity for bodily decoration, it is surely likely that they could exchange a word or

two with each other. Maybe they were speaking well before *Homo sapiens*.

6. We can clearly imagine a tactile form of communication in which "speech acts" are performed by touching the recipient in specific ways. Ants touch each other in order to communicate, as with "antennae tapping." In principle, any sense could be the receptor for acts of communication.

7. Heidegger, *Being and Time*. Although he speaks a lot of the "ready-to-hand," Heidegger never really places any emphasis on the hand as such; he just discourses generally about instrumentality in relation to human *Dasein*.

8. A very naïve view of how language evolved would propose that we had lots of interesting and profound thoughts that we felt compelled to share with others, so we just began blurting them out one day, as a form of entertainment. The more realistic view is that language developed in a context of biological necessity and was based on anatomical and cognitive structures that had evolved earlier for other purposes, specifically tool use and hand development, along with social cooperation. We learned to communicate because we needed to in order to survive, not as a luxury. The origin of language must have been surrounded by fear and anxiety, as well as physical labor, not by a desire to have stimulating conversations. Language was born in conditions of biological catastrophe, with extinction always looming. It was an act of desperation, not aspiration (as if our ancestors thought, "What a fine thing it would be to be able to speak"). I say all this because it is easy now to think of human language as a source of pleasure and culture, as a sort of gift from on high. But the more realistic picture is that language was forced on us by the tragedy of being driven from our ancient arboreal homeland into alien and threatening new territory. We *resorted* to language in order to cope with our new circumstances as deracinated refugees.

9   Hand and Mind

1. Other forms of advanced intelligence are conceivable. Some alien species elsewhere in the universe may have no talent for tools but be

very intelligent at philosophy or mathematics or psychology (though it may be hard to tell a convincing story about how they came to be so). It may even be that our specifically tool-oriented intelligence, with its particular origins and purposes, ill equips us for some areas of intellectual inquiry. Tools may have liberated our brain in one direction but inhibited it in other directions. Ecological niche always constrains brain function. We have the brain of an artisan, basically; but not every conceivable form of intelligence has to be of this limited type.

2. Jean Piaget's many studies of intellectual maturation in the child make a lot of the role of sensorimotor activity, and such activity centers on the hand.

3. On enactive theories, see Alva Noë, *Action in Perception*.

4. Is it an accident that G. E. Moore held up his *hands* while endeavoring to prove the existence of the external world, in his "A Defence of Common Sense"? Our hands are certainly very real to us.

5. The accuracy of perception by the hand is obviously very important for survival: you can't afford to be wrong about what you are gripping (snake or spear). Delicate manual perception is also advantageous. Things in the hand can easily damage the hand. It is not so with the eye: you can misperceive a thing and the eye is unlikely to be damaged. Touching sharp pointy objects is a different matter from seeing them. We have to be careful with our hands, as we employ them perceptually.

6. Obviously, I am being a little simplistic here: vision also plays a major role in determining how we conceive of things. The empiricist tradition enshrines this primacy of vision. I am trying to find a place for something that tradition ignores: the role of the hand in shaping our conception of the world. Gripping and touching up close plays at least as large a role as seeing from a distance in forming our concepts of material things, I contend. The idea of solidity surely comes from touch, not vision, and our egocentric space is a haptic space above all.

7. A canine psychoanalyst might detect anxiety centering on the jaws and teeth: impotent jaws that have lost all prehensile power, crumbling teeth, muzzles. A dog's anxiety dreams likely concern catastrophes of the mouth area, where its prehensive powers are concentrated.

8. The philosophical problems contemplated by a specific type of species might vary according to its own makeup. Abstract beings (if such there could be) would be troubled by the concrete; disembodied beings (ditto) would be perplexed by the notion of a body; divine beings (again ditto) might be baffled by evil and weakness of will; beings made of Platonic universals (double ditto) might find the notion of a particular unintelligible; beings with both minds and bodies might be puzzled about how their two sides interact. (Just to be clear: I do not take these possibilities very seriously as sober metaphysics—I am just trying to make a point about the role of the body in shaping one's sense of things. The same applies to the previous note.)

9. See Broad, "Some Elementary Reflections on Sense Perception."

10. Here is a linguistic oddity: we can say "*x* grasps that *p*" but we can't say "*x* grips that *p*"—"grips" only allows the direct-object form. We have the same contrast with "holds" and "seizes." I am not sure why this is.

11. If we want a slogan, we could try "psychology recapitulates anatomy"—a variant on Freud's "anatomy is destiny." More cumbersomely, we could adopt "psycho-type mirrors phenotype." Or we could go very old-fashioned with "the soul is shaped by the body."

10   Selective Cognition and the Mouth

1. There is a general problem of terminology (and conceptualization) here, because our psychological vocabulary evolved mainly to describe our own psychology, not that of other species, present or past. Our terms and concepts are therefore parochial and fail to fit neatly with the minds of other species. But this does not mean that other species lack a determinate psychology—only that our means of description are inadequate. The *variety* of minds is not well captured by our anthropocentric psychological vocabulary. This applies *a fortiori* to ancient species long extinct.

2. Methodologically, the argument goes from the abstract morphology of the two capacities to a thesis of derivation—the *reason* for the coincidence of form is that one came from the other. Thus there is no salta-

tion. It is just like arguing that hands and feet came from fins because of their morphological resemblance. Note that this is not deduction but inference to the best explanation: the best explanation of the similarity is derivation, though this does not of course logically follow from the similarity. Derivation is a hypothesis about the reason for the similarity.

3. The notion of being "subject to the will" in application to thought is explored in McGinn, *Mindsight*, chap. 1. We are trying to explain (*inter alia*) the transition from passive perceiver to active thinker, with active oral prehension playing a key role.

4. A more familiar example of exaptation is provided by bird feathers: feathers originally evolved as a means of thermal regulation and only subsequently were co-opted for flight. And notice that in this case the underlying brain machinery is co-opted too, because the wings need instructions from the brain to induce them to open, close, and flap when performing either function.

5. Once a trait evolves it can become detached from its earlier function, so that selective cognition might become quite independent of oral prehension. Oral prehension could even wither away, leaving behind the new trait that evolved from it long ago. But perhaps for the earliest possessors of selective cognition there was always an association with the mouth—thoughts of an object were always accompanied by oral sensations of the kind involved in holding an object in the mouth. Maybe in an intermediate species selective cognition was always accompanied by salivation! It would be fascinating to discover that concentrating on an object in thought is regularly associated with an imperceptible clenching of the jaw.

6. I came up with this theory by asking what bodily trait was more or less coextensive with cognition in the animal kingdom, and then conjecturing that this trait was the preadaptation for cognition. I reasoned that cognition is extremely widespread, though not universal, and the bodily correlate seemed to be advanced oral prehension. Then I noticed the abstract similarities between the two—hence the theory.

7. In the pop song "Hold Me Tight" (written and performed memorably by the Beatles) the sentiment is not "Apply X pounds of pressure to my body" but something a lot more personal, involving the action of one agent on another. Similarly, when we say a person has grabbed something, we don't just mean that his hand has abruptly clamped around it.

8. See McGinn, *Mindsight*, chapter 1, for a discussion of the general concept of the visual.

9. When Frege speaks of grasping a sense, one gets the idea of the mind curling around the sense, as if encircling it in thought, apprehending its inner structure. The sense is an articulated entity, according to Frege, and the mind's grasp of it mirrors this articulation—as the shape of the grasping hand mirrors the shape of the object grasped. Thus the mind acts on the sense and the sense acts back on the mind, just as with grasping an object in the hand. That, at any rate, is how one is apt to envisage mental grasping.

10. To say that bird flight evolved from thermal regulation by means of feathers is not to say that flight is *reducible* to thermal regulation. It is to say rather that the mechanism of flight, viz. feathers, had its precursor in feathers used to regulate temperature. Feathers served both functions, as the brain circuits underlying oral prehension might also serve selective cognition (suitably supplemented).

11   The Origin of Sentience

1. Panpsychism is an old view and perennially attractive. I don't care for it myself, but for expository purposes I shall take it as given. For one defense see Nagel, "Panpsychism," in his *Mortal Questions*.

2. As is often noted, the brain is metabolically costly, so sentience will be too, since it needs brain tissue in abundance. So sentience will not evolve unless there is a pressing need for it. My question is what the *pressing* need is for sentience among organisms on Earth: for it cannot be a mere luxury. Sentience is a costly adventure in high upkeep technology, liable to drain the organism's energy resources. Plants and microorganisms don't bother with it, so it is not essential to life as such—so why do so many animals take on such a costly burden?

3. This is what happens with the tails of many lizards: one tail gets bitten off but another one grows to replace it. This is quite a costly adaptation in terms of lizard hardware, but it is presumably worth the price (lizards with it do better than lizards without it). Being able to grow another head on demand would be equally adaptive, but we must assume that that is beyond the technology available to lizard genes.

4. I don't mean logically sufficient, because predator detection *could* occur without real sentience—it could be wholly robotic. But it is sufficient in the sense that evolution on Earth seems to have chosen this particular way of registering information. I doubt there are any higher organisms equipped with predator-detecting senses that are purely robotic on Earth (their brains are all made of much the same material). So we must assume that on Earth, sentience is the optimal way to register information about the environment.

5. This means that the existence of highly developed brains and consciousness itself traces back to predation. Everything we value in life thus depends on the existence of predation—no predation, no mind in any meaningful sense. In a way, then, everything good has its origin in something abhorrent—including the moral sense that rebels at predation. It was only when predators troubled our heretofore peaceful planet that sentience was born—and hence brought about the products of consciousness. We are conscious because we might be eaten; we think because we might become food for big cats. Those feared predators are what lifted us to the level of conscious beings.

6. By contrast, I think there are clear psychological relics of our arboreal past, as I suggest in chap. 14.

7. This is hard to verify, of course. I suppose we could check to see if the brain of the fetus goes through a stage resembling the brain of a fish. It is quite true that the fetus must have oceanic sensations as it floats in the amniotic fluid, but this may not be much like the experience of your typical fish. At any rate, it is nice to imagine that I once knew what it is like to be a fish—in the embryonic days before my brain matured to mammalian proportions. (What if the fetus went through a bat-brained stage?)

8. Suppose we could compute a "prehensivity index" that enabled us to measure and quantify a given animal's degree of prehensive accomplishment, so that a ranking of species could be established. It seems a reasonable conjecture that there would be a lawlike correlation between the prehensivity index and an animal's degree of psychological sophistication. We would then have a very general "psychophysical law" relating one magnitude to another: maybe intelligence would be a logarithmic function of prehensive power, as with some laws of psychophysics. I suspect we humans would come out highest on both measures, mainly because of our abnormally prehensive hands. But none of this would be hard science, needless to say.

12   The Meaning of the Grip

1. Pushing the thought experiment through requires prescinding from the kind of feet we have and from breathing air and even from the way the body envelope keeps a grip on the inner organs—for these are all (quasi-)prehensive matters. To imagine oneself as comprehensively non-prehensive is no easy feat. Still, the *idea* of it can be contemplated.

2. There is thus a deep difference between merely perceiving an object, say visually, and actually gripping it in one's hand: the perceiving is not itself de-alienating, just revealing of an objective plane of being, but the gripping is an active incorporating engagement (people get "engaged"). It is a hitching together, a physical fusion, an act of synthesis; but seeing by itself is still remote and unengaged—a mere positing. Gripping is warm intimacy; seeing is cool detachment. (Smell is somewhere between the two, because it is a distance sense that involves holding emanations from the object in one's nose.) There is therefore something a little off about describing sense perception as "prehension"—almost, but not quite.

3. The one part of the body we cannot grip is the gripping hand itself—a hand cannot grip itself as it grips. I suppose the hand can be said to be able to get a partial grip on itself, by pressing the fingers into the palm or clasping the index finger with the thumb, but it cannot grip itself as it grips other objects (try gripping the back of your hand with the fin-

gers of that hand). The hand is the ungripped gripper. But this prehensive isolation is mitigated by the fact that one hand can grip the other, and this simple act brings prehension of the body full circle. It would be odd not to know what it feels like to grip one's own hand—as must be the case for those with only one hand. I imagine it is a funny feeling, as if part of one's body is slightly "other."

4. Sartre, *Being and Nothingness*. His famous example of bad faith involves a woman's hand: she renders it inert and thinglike as her suitor hopefully clasps it. She has tried to transfer it from the domain of the for-itself into the domain of the in-itself, as if her hand is not *free*. (The hands are free agents *par excellence*.)

5. Hume described causation as "the cement of the universe": the grip is the cement of the *social* universe. And isn't causation a "grippy" matter itself? It holds the universe together, in Hume's image—the causing object reaches out to the effect object and seizes it. Here we see the glimmerings of "prehensional metaphysics," whereby the universe is conceived as one big theater of prehension, even down to elementary particles. Just such a metaphysical system was propounded by A. N. Whitehead in works such as *The Concept of Nature*. I don't myself subscribe to prehensional metaphysics, despite my enthusiasm for prehension, because I think only organic entities can literally grip or grasp: but I do appreciate the motivation.

6. Even a "harmless" tool, such as a paper clip or a book, can prove deadly in certain circumstances: the clip can be swallowed, the book used as a cudgel. Kitchen knives are clearly dangerous implements, to be used with care. Even the innocuous teaspoon can be used to gouge out an eye. These objects all perform a job for us, like obedient servants, but they can turn against us in a flash. They are for us but they are also against us. Strangling with a shoelace provokes a peculiar shudder of recognition.

7. The entire digestive tract is a grip-release mechanism: swallowing is an act of extreme gripping (like Jonah and the whale). After digestion, the body holds on to the derived nutrients, while letting go of the waste.

8. Two of the main proponents of this school of child psychology are Melanie Klein and John Bowlby. Interested readers might wish to consult Klein's *The Psychoanalysis of Children* and Bowlby's *Attachment and Loss*—though my use of their ideas requires no detailed knowledge of their theories. My summary of their views in the text is intended only to provide a sketch of the relevant material, so as to illustrate the role of prehension in interpersonal relations.

9. We might speculate that humans (and other animals) have an innate prehension program designed to establish emotional relations over the course of maturation—and this program must not be thwarted. The notorious work by Harry Harlow and associates on maternal deprivation in rhesus monkeys illustrates the importance of physical attachment to affective attachment. But I cite this work reluctantly, because of the complete lack of ethical restraint involved in these obviously cruel experiments.

10. Educating people in the use of tools and implements of all kinds should emphasize the importance of the grip involved. The pleasures of the grip should also be highlighted, so that it doesn't seem so much like work—tell people to take a moment to *feel* the object in their hand. In tennis, say, it pays to attend to the precise way the racquet is gripped, especially because of the necessity to switch grips without looking, as well as for other reasons (Eastern grip, Western grip, semi-Western grip, etc.).

11. In these reflections I am influenced by Richard Dawkins's *The Extended Phenotype*, in which he argues that the beaver's genes as much build dams as they build beavers. The dam is really part of the beaver's (extended) phenotype. You might protest that dams and some of the items on my list do not *grow* from the organism concerned: thus beards are part of the body but spectacles are not. True enough, but other examples show that growing from the body is not a necessary condition for being considered part of it—such as prosthetic organs and limbs, as well as organic transplants. Functional connectedness is what counts, not "growing from." A transplanted kidney did not grow from its current owner but it is surely part of his or her body. Nor is the condition sufficient, or else your cut hair would still be part of your body. If you

discovered that your heart is actually a parasitic organism in symbiotic relation to you, would you conclude that it is not part of your body? What does the job of a body part *is* a body part, roughly. The important point is that human tools play the same kind of role in human survival as human limbs or teeth or eyes—they are useful adaptations subject to natural selection.

13   A Culture of Hands

1. If our bodies were built according to the proportions of a cortical homunculus diagram, the hands would be at least the size of the rest of the body, not counting the grotesquely bloated lips and tongue—because so much of the brain is devoted to them. The hands are anatomically small but cortically large, owing to their great functional complexity (contrast the buttocks).

2. I advocate emphasizing the biological naturalness of writing in our schools. Writing is an aspect of our evolutionary heritage, a link to our ancestors, and a major factor in our species ascendancy. Writing is in our genes, in our blood—because speaking with the hands is (everyone gestures). I vaguely remember learning to write and finding it an absorbing and astonishing accomplishment (and I treasured fountain pens). Even today I love the feeling of a pen in my hand as I rapidly inscribe words and sentences on paper—so forceful yet so delicate, and so full of meaning! Writing is the perfect fusion of movement, tool, and thought. Steven Pinker should write a book called *The Writing Instinct*. Yes, we have to learn to do it, but it harks back to primitive manual abilities, linking thoughts with hands, cognition with prehension. For some people, concentrated thought is not possible without writing.

3. The connection between hands and disgust could use extended treatment. I neglected the subject in my book *The Meaning of Disgust*. The hands seem the focal point of our disgust reactions, second only to the mouth. Grasping the disgusting thing, with hand or mouth, feels especially loathsome, as opposed to simply perceiving it or even touching it with other parts of the body (say, the elbow). We especially don't like disgusting material clinging to our fingers. Compulsive hand washing is

part of this general repugnance. When, I wonder, did early man develop these fastidious feelings about his hands?

4. I mean that the hand demonstrates the remarkable creative power of evolution by natural selection. If natural selection can produce such an astonishing organ, what can it not produce? This should make it easier to accept that natural selection led to art, science, philosophy, morality, and so on—these being no more amazing or "above nature" than the hand (which made all these accomplishments possible anyway). The human brain, remember, is largely a by-product of the hand, coevolving with it in symphonic harmony (each conducting the other).

14   Arboreal Remnants

1. Just ask yourself what early humans would think if confronted by the accomplishments of their descendants. Do you think they would say: "Yes, just as I thought—we have become nature's top guns"? No, they would be astonished at the future state of man: his technology, his life span, and his sheer numbers. In hindsight, perhaps, we can discern the seeds of future success, but primitive stone tools and a few hand gestures don't look much like the stuff of world domination. Why aren't the lions and tigers running the world?

2. You might suppose that we could just use our natural tool-using intelligence to make new tools and rebuild our world, but what if that part of the brain ceased to function and we no longer had tool-using abilities? Our big brain—good for science, philosophy, culture, and such—would not be much help in the struggle for survival without the use of tools. Profound thoughts do not keep you safe and fed. And matters would turn even more desperate if our hands became stiff and clumsy. We are lucky that no viruses target the hands. Dominant as we are, we are also inches away from wholesale destruction. We just *happen* to be good survivors (so far).

3. We can hold our hands above our head, we have a rotating shoulder girdle, a strong grip, and we have the right musculature on the back and arms. Of course, none of this is surprising, given that we are descended from apes and ancestral preservation is the rule.

4. As a psychological experiment, I suggest watching some film of gibbons brachiating and seeing whether you feel the urge to join them in the trees. As a physical experiment, we could train people in brachiation skills and find out whether they improve physically and mentally, compared to a control group. (Since writing the above about brachiation and the fitness industry, I have discovered a startup company called "Darwinian Fitness" that advocates brachiation and other tree-related exercises as better alternatives to standard muscle-building routines, based precisely on evolutionary thinking.)

5. Napier writes as follows: "Man is a product of a primary arboreal background and a secondary ground-living heritage. He possesses the 'flight' responses of forest living monkeys and the 'fight' responses of the ground-living baboons and macaques. His genetic ambivalence confronts him every moment of the day. Do I get in there and fight, or do I settle for what I have? Do I 'twist' or 'buy'? Our uncertainty, the racking moments of self-doubt teetering on the precipice of indecision is a reflection, not of our ambivalent present, but of our ancient mixed-up past" (*The Roots of Mankind*, 220).

6. Some contemporary humans, though not many, still make their homes in trees: thus the Korowai and Kombai tribes of Papua New Guinea, who live in tree houses often very high above the ground. From all reports, they are quite happy with this domestic arrangement. Quite a few people living in the suburbs build tree houses in their yards (there is even a TV series about such people).

7. In some dystopian fantasies, humans are compelled to live underground by a catastrophe at the Earth's surface, say nuclear devastation or total pollution. This is descending one step further down the ladder from the trees, and an even gloomier prospect. We certainly don't want to live like *moles* (though moles don't seem to mind)—as if banished completely from our arboreal past. We are climbers and swingers, not burrowers, or even browsers.

8. The domestic cat is an interesting point of comparison. Though a typical house will lack any indoor trees, cats will readily climb whatever is available—evidently retaining urges from their semiarboreal days.

They are still physically equipped for climbing and psychologically they are inclined to it. We humans also display vestigial climbing urges, though tree climbing is a minority activity. People like to climb hills and mountains, stairs and steps, professional and social hierarchies. We are still partly a climbing species (unlike the elephant), despite our latter-day domestication on the flatlands. The desire to ascend still throbs within us, as a remnant of our professional tree-climbing days.

9. One of the primary cognitive skills required by any animal is the ability to recognize conspecifics, which involves recognizing members of other species as *not* of one's own species. This is of particular importance when it comes to predator avoidance. It is probably innate in most if not all species. Humans are no different: we too have an innate ability to distinguish species, and hence harbor a deep sense of species similarity. Thus the recognition of our kinship with birds rests on an innate and heritable foundation. Perhaps our general fondness for birds reflects this inbuilt recognition of kinship. I suspect we have an unspoken ambivalence about eating birds in general because of this, especially if we view the species in question as quite similar to us (for some reason we make an exception for chickens). Eating *talking* birds feels a lot like cannibalism; and the Bird Man of Alcatraz could hardly *eat* his bird companions.

10. We could have become ground-dwelling quadrupeds after our descent, using our old gripping hands as mere feet, with the fingers losing their prehensile facility—which is what appears to have happened to some other primates that came down from the trees. But that was not a good route for us, because it was not an optimal use of our anatomical strengths—the hand was far too good to waste as a mere foot. The following counterfactual seems true: if we had become quadrupeds, not bipeds, we would now be extinct—and for a long time.

11. Napier tells us: "In the sense of the number of hairs per square centimeter of skin, man is as hairy as a gorilla but his hairs are so fine and colorless as to be almost invisible over most of his body" (*The Roots of Mankind*, 143–144). I take his expert word for it.

12. On the "scavenging hypothesis," see McKee and Poirier, *Understanding Human Evolution*, 208.

13. Though even shopping can turn ugly when shoppers feel competition from other shoppers for bargains. This resembles nothing so much as scavengers converging on a nice juicy corpse and fighting each other for the best bits.

14. The fundamental reason for psychological remnants is anatomical ancestral preservation in the brain. The brain is the basis of the mind, and it is not totally scrapped and rebuilt when adaptation or speciation occurs; rather, it is modified and supplemented, with old structures left intact. The human brain does not relate to ancestral brains as a brand new building relates to the previous one that was totally demolished on the same site. It relates rather as a modified building does to a previous building—with the old cellars and outhouses left intact, though no longer used. In short: old brains never die, they just fade into the background.

15 The Future of the Hand

1. This is equivalent to the question of why the human brain became so large and complex. Why don't other species have a brain as large and complex as ours, particularly other primates? Such a brain would be useful (though metabolically costly), but other animals have not evolved it. So: what did our ancestors have *prior* to encephalization that *enabled* encephalization? What was the impetus for the human cerebral cortex?

2. Pinker takes much the same line in *How the Mind Works*, in the aptly named chapter "Revenge of the Nerds" (see 197–198). Here is my wild estimate of the probability of intelligent life elsewhere in the universe, as opposed to less nerdy forms of life: if there are a million planets in the universe with advanced life, even life as sophisticated as primates, then there are no more than *three* with intelligent life. Nerds only emerge in exceedingly rare conditions (which is why we are the only nerd species on Earth).

3. Two factors clearly militate against developing *Intelligence*: (a) it needs a metabolically costly bulked-up brain that devours energy; (b) a brain that size needs a big skull to contain it, and this produces prob-

lems in giving birth because of the relative diameter of the birth canal. Given these two drawbacks, it is not surprising that the vast majority of species have been content with a more modest brain and hence less *Intelligence*. And it is entirely possible to thrive and multiply with a smaller head organ and nothing in the way of tools or language (consider insects). As an adaptation, *Intelligence* is just not that cool—despite being otherwise admirable. We tend to overestimate its adaptive value from our limited anthropocentric perspective. It has its great adaptive advantages, to be sure, but also some notable drawbacks.

4. Darwin comments: "It might have been an immense advantage to man to have sprung from some comparatively weak creature," adding: "The slight corporeal strength of man, his little speed, his want of natural weapons, &c., are more than counterbalanced, firstly by his intellectual powers, through which he has, while still remaining in a barbarous state, formed for himself weapons, tools, &c., and secondly by his social qualities which lead him to give aid to his fellow-men and to receive it in return" (*The Descent of Man*, 93–94).

5. Imagine if the anti-hand zealots insisted on *binding* the hands of children, as the Chinese used to bind the feet of girls, thus rendering them deformed and dysfunctional. These ruined hands could never play an instrument, write a sentence, caress a loved one, play tennis, or do anything else manual. Doesn't that sound like a truly terrible state of affairs?

Bibliography

Armstrong, David F., William C. Stokoe, and Sherman E. Wilcox. *Gesture and the Nature of Language*. Cambridge: Cambridge University Press, 1995.

Bell, Charles. *The Hand: Its Mechanism and Vital Endowments as Evincing Design*. Cambridge: Cambridge University Press, 2009. (Original work published 1833.)

Bowlby, John. *Attachment and Loss*, vol. 1: *Attachment*, 2nd ed. New York: Basic Books, 1999. (Original work published 1969.)

Bowlby, John. *Attachment and Loss*, vol. 2: *Separation: Anxiety and Anger*. London: Hogarth Press, 1973.

Bowlby, John. *Attachment and Loss*, vol. 3: *Loss: Sadness and Depression*. London: Hogarth Press, 1980.

Broad, C. D. "Some Elementary Reflections on Sense Perception." In *Perceiving, Sensing, and Knowing*, ed. R. J. Schwartz. Berkeley, CA: University of California Press, 1965.

Darwin, Charles. *The Descent of Man, and Selection in Relation to Sex*. New York: Dover, 2010. (Original work published 1871.)

Dawkins, Richard. *The Ancestor's Tale: A Pilgrimage to the Dawn of Evolution*. New York: Houghton Mifflin, 2004.

Dawkins, Richard. *The Extended Phenotype: The Long Reach of the Gene*. Oxford: Oxford University Press, 1989.

d'Errico, Francesco, and Lucinda Backwell, eds. *From Tools to Symbols: From Early Hominids to Modern Humans*. Johannesburg: Witwatersrand University Press, 2006.

Gibson, Kathleen R., and Tim Ingold, eds. *Tools, Language, and Cognition in Human Evolution*. Cambridge: Cambridge University Press, 1993.

Heidegger, Martin. *Being and Time*. Trans. John MacQuarrie and Edward Robinson. London: SCM Press, 1962.

Johanson, Donald, and Blake Edgar. *From Lucy to Language*. New York: Simon & Schuster, 1996.

Klein, Melanie. *The Psychoanalysis of Children*. Charleston, SC: Nabu Press. (Original work published 1923.)

McGinn, Colin. *Mindsight: Image, Dream, Meaning*. Cambridge, MA: Harvard University Press, 2006.

McGinn, Colin. *The Meaning of Disgust*. Oxford: Oxford University press, 2011.

McKee, Jeffrey K., Frank E. Poirier, and W. Scott McGraw. *Understanding Human Evolution*, 5th ed. Upper Saddle River, NJ: Pearson, 2004.

McNeill, David. *Hand and Mind: What Gestures Reveal about Thought*. Chicago: University of Chicago Press, 1992.

Moore, G. E. "A Defence of Common Sense." In G. E. Moore, *Philosophical Papers*. London: Routledge, 1959. (Original work published 1925.)

Nagel, Thomas. *Mortal Questions*. Cambridge: Cambridge University Press, 1979.

Napier, John. *Hands*. Princeton: Princeton University Press, 1980.

Napier, John. *The Roots of Mankind*. Washington, DC: Smithsonian Institution Press, 1970.

Noë, Alva. *Action in Perception*. Cambridge, MA: MIT Press, 2006.

Pinker, Steven. *How the Mind Works*. New York: W. W. Norton, 1997.

Radman, Zdravko, ed. *The Hand, an Organ of the Mind: What the Manual Tells the Mental*. Cambridge, MA: MIT Press, 2013.

Sartre, Jean-Paul. *Being and Nothingness*. New York: Washington Square Press, 1992. (Original work published 1943.)

Tallis, Raymond. *The Hand: A Philosophical Inquiry into Human Being*. Edinburgh: Edinburgh University Press, 2003.

Whitehead, A. N. *The Concept of Nature*. New York: Cosimo, 2007. (Original work published 1920.)

Wilson, Frank R. *The Hand: How Its Use Shapes the Brain, Language, and Human Culture*. New York: Vintage, 1998.

Wittgenstein, Ludwig. *Tractatus Logico-Philosophicus*. Trans. D. F. Pears and B. F. McGuinness. London: Routledge & Kegan Paul. (Original work published 1922.)

Index